BioWatch PCR Assays

Building Confidence, Ensuring Reliability

Committee on PCR Standards for the BioWatch Program

Board on Life Sciences
Division on Earth and Life Studies

Board on Health Sciences Policy
Institute of Medicine

NATIONAL RESEARCH COUNCIL *AND*
INSTITUTE OF MEDICINE
OF THE NATIONAL ACADEMIES

Abbreviated Version

THE NATIONAL ACADEMIES PRESS
Washington, D.C.
www.nap.edu

THE NATIONAL ACADEMIES PRESS 500 Fifth Street, NW Washington, DC 20001

NOTICE: The project that is the subject of this report was approved by the Governing Board of the National Research Council, whose members are drawn from the councils of the National Academy of Sciences, the National Academy of Engineering, and the Institute of Medicine. The members of the committee responsible for the report were chosen for their special competences and with regard for appropriate balance.

This study was supported by Contract No. HSHQDC-11-D-00009/Order No. HSHQDC-13-J-00274 between the National Academy of Sciences and the U.S. Department of Homeland Security. Any opinions, findings, conclusions, or recommendations expressed in this publication are those of the author(s) and do not necessarily reflect the views of the organizations or agencies that provided support for the project.

International Standard Book Number-13: 978-0-309-36722-6
International Standard Book Number-10: 0-309-36722-0

Additional copies of this report are available for sale from the National Academies Press, 500 Fifth Street, NW, Keck 360, Washington, DC 20001; (800) 624-6242 or (202) 334-3313; http://www.nap.edu.

Printed in the United States of America.

Suggested citation: NRC (National Research Council) and IOM (Institute of Medicine). 2015. *BioWatch PCR Assays: Building Confidence, Ensuring Reliability.* Washington, DC: The National Academies Press.

THE NATIONAL ACADEMIES

Advisers to the Nation on Science, Engineering, and Medicine

The **National Academy of Sciences** is a private, nonprofit, self-perpetuating society of distinguished scholars engaged in scientific and engineering research, dedicated to the furtherance of science and technology and to their use for the general welfare. Upon the authority of the charter granted to it by the Congress in 1863, the Academy has a mandate that requires it to advise the federal government on scientific and technical matters. Dr. Ralph J. Cicerone is president of the National Academy of Sciences.

The **National Academy of Engineering** was established in 1964, under the charter of the National Academy of Sciences, as a parallel organization of outstanding engineers. It is autonomous in its administration and in the selection of its members, sharing with the National Academy of Sciences the responsibility for advising the federal government. The National Academy of Engineering also sponsors engineering programs aimed at meeting national needs, encourages education and research, and recognizes the superior achievements of engineers. Dr. C. D. Mote, Jr., is president of the National Academy of Engineering.

The **Institute of Medicine** was established in 1970 by the National Academy of Sciences to secure the services of eminent members of appropriate professions in the examination of policy matters pertaining to the health of the public. The Institute acts under the responsibility given to the National Academy of Sciences by its congressional charter to be an adviser to the federal government and, upon its own initiative, to identify issues of medical care, research, and education. Dr. Victor J. Dzau is president of the Institute of Medicine.

The **National Research Council** was organized by the National Academy of Sciences in 1916 to associate the broad community of science and technology with the Academy's purposes of furthering knowledge and advising the federal government. Functioning in accordance with general policies determined by the Academy, the Council has become the principal operating agency of both the National Academy of Sciences and the National Academy of Engineering in providing services to the government, the public, and the scientific and engineering communities. The Council is administered jointly by both Academies and the Institute of Medicine. Dr. Ralph J. Cicerone and Dr. C. D. Mote, Jr., are chair and vice chair, respectively, of the National Research Council.

www.national-academies.org

COMMITTEE ON PCR STANDARDS FOR THE BIOWATCH PROGRAM

GEORGES C. BENJAMIN (*Co-Chair*),[†] Executive Director, American Public Health Association, Washington, DC

KENNETH I. BERNS (*Co-Chair*), Distinguished Professor of Molecular Genetics and Microbiology Emeritus, University of Florida College of Medicine, Gainesville

EDWIN P. PRZYBYLOWICZ (*Co-Chair*),[‡] Senior Vice President and Director of Research (retired), Eastman Kodak Company, Webster, NY

BRUCE BUDOWLE, Executive Director, Institute of Applied Genetics, and Professor, Department of Molecular and Medical Genetics, University of North Texas Health Science Center, Fort Worth

CHARLES CHIU, Associate Professor, Laboratory Medicine and Medicine, Infectious Diseases; Director, UCSF-Abbott Viral Diagnostics and Discovery Center at China Basin; and Associate Director of the UCSF Clinical Microbiology Laboratory, University of California, San Francisco

JOHN M. HARDHAM, Associate Research Fellow and Chair, Emerging Infectious Disease Program, Zoetis, Inc., Kalamazoo, MI

GRACE KUBIN, Director, Laboratory Services Section, Texas Department of State Health Services, Austin

M. ALLEN NORTHRUP, Principal, Northrup Consulting Group, San Francisco, CA

TOM SLEZAK, Program Leader, Global Security Program, E.O. Lawrence Livermore National Laboratory, Livermore, CA

PETER M. VALLONE, Leader, Applied Genetics Group, Biomolecular Measurement Division, National Institute of Standards and Technology, Gaithersburg, MD

Study Staff

KATHERINE BOWMAN, Study Director, Board on Life Sciences

AUTUMN DOWNEY, Program Officer, Board on Health Sciences Policy

KATHRYN HUGHES, Senior Program Officer, Board on Chemical Sciences and Technology

JON Q. SANDERS, Program Coordinator, Board on the Health of Select Populations (until November 2014)

NIA D. JOHNSON, Senior Research Associate, Board on Army Science and Technology (from November 2014 to March 2015)

[†] Since June 2014.

[‡] Until May 2014.

CLAIRE BALLWEG, Senior Program Assistant, Board on Environmental Studies and Toxicology (from May 2015)
BETHELHEM MEKASHA, Financial Associate, Board on Life Sciences
ANDREW POPE, Director, Board on Health Sciences Policy
FRANCES SHARPLES, Director, Board on Life Sciences

BOARD ON LIFE SCIENCES

BOARD ON HEALTH SCIENCES POLICY

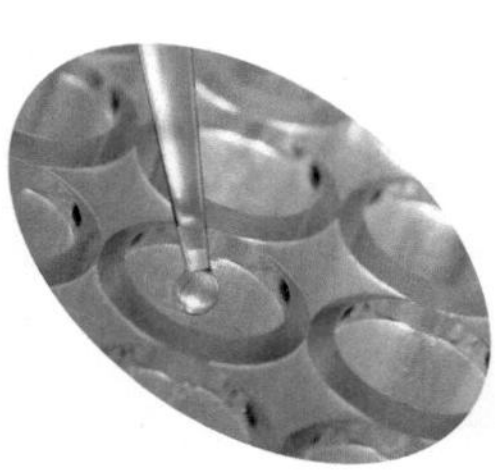

Acknowledgments

This report has been reviewed in draft form by individuals chosen for their diverse perspectives and technical expertise, in accordance with procedures approved by the National Research Council's Report Review Committee. The purpose of this independent review is to provide candid and critical comments that will assist the institution in making its published report as sound as possible and to ensure that the report meets institutional standards for objectivity, evidence, and responsiveness to the study charge. The review comments and draft manuscript remain confidential to protect the integrity of the deliberative process. We wish to thank the following individuals for their review of this report:

Cynthia Bruckner-Lea, *Pacific Northwest National Laboratory*
Nancy D. Connell, *Rutgers New Jersey Medical School*
Christina Egan, *New York State Department of Health*
Sean Eddy, *Howard Hughes Medical Institute Janelia Farm Research Campus*
Joan Gebhardt, *Naval Medical Research Center*
Jonna A.K. Mazet, *University of California, Davis*
Suzet M. McKinney, *Chicago Department of Public Health*
R. Paul Schaudies, *GenArraytion, Inc.*
Victor Waddell, *Arizona Department of Health Services*

Although the reviewers listed above have provided many constructive comments and suggestions, they were not asked to endorse the conclusions or recommendations nor did they see the final draft of the report before its release. The review of this report was overseen by **Ronald M. Atlas**, *University of Louisville* and **David R. Walt**, *Tufts University*. Appointed by the National Research Council, they were responsible for making certain that an independent examination of this report was carried out in accordance with institutional procedures and that all review comments were carefully considered. Responsibility for the final content of this report rests entirely with the authoring committee and the institution.

The committee thanks all those who provided information during the course of the study. The committee particularly thanks Kerby Shedden, University of Michigan, and Michael Cohen, National Research Council, for their assistance and thoughtful conversations on statistical confidence intervals and experimental design.

Preface

Knowing when a health threat enters a community is the cornerstone of biopreparedness. Historically, public health officials have had to rely on a clinical detection system that includes the astute clinician, the unanticipated lab result, or a cluster of disease cases to know that some disease threat was present. These methods of detection, while important, may put the health security system behind the curve in responding to a disease threat in the case of a bioterrorist attack. The BioWatch program is designed to get upstream of these clinical detection systems by providing early warning to local, state, and federal partners should an airborne release of certain types of biothreat agents occur. The program relies on collectors deployed in jurisdictions around the country, and on sample results provided through public health laboratories. It is a component of a much broader monitoring and surveillance effort to identify and respond to terrorist threats that ensures the health security of the nation.

Ensuring health security requires an early warning system with which its users have a high degree of confidence. The users need a clear understanding of the meaning and limitations of the data they receive and have trust that they can use the results to make subsequent, potentially high-consequence decisions. As the report discusses, standards and validation provide a necessary basis for confidence in the system.

Although standards may not seem glamorous, they are fundamental. Our own experiences studying infectious microorganisms and working to support a strong and effective public health system lent us motivation for the task. The committee welcomed the opportunity to bring attention to the critical roles that standards play in supporting a program such as BioWatch, the efforts that have taken place to develop performance standards that could be applicable to program needs, and the base of information that has been gathered on BioWatch assay and system performance. Based on our statement of task, the report focuses on performance standards for the polymerase chain reaction (PCR) assays used in BioWatch, while emphasizing that all of the system's components are necessarily interconnected. It is evident that the rapid development of next-generation sequencing offers the promise of a more definitive approach to the secondary verification or confirmation step required after the initial PCR assay in order to secure the trust required for responsible public health officials and other government authorities to initiate an appropriate response to a select agent detection.

We were joined in our task by fellow committee members with diverse areas of expertise in genetics and molecular biology, bioinformatics, technology development, and public health. We thank them for their commitment to the study topic and for giving so willingly of their time and energy. It has been a pleasure to work with each of them. The report was also informed by presentations and discussions with a number of experts in environmental biosurveillance, pathogen biology, biodetection, assay and standards development, and validation from within and outside the federal government, as well as from state and local BioWatch jurisdictions. Their contributions were critical in informing the committee and we thank them for their willingness to share their views with us.

Finally, on behalf of the committee, we would like to thank the staff of the National Research Council and the Institute of Medicine for their assistance and dedication in working alongside us. Their support was integral in bringing the project to fruition.

Georges C. Benjamin
Kenneth I. Berns
Co-Chairs

Appendix A to the Committee's report has been determined by the sponsor to contain information described in one or more of the exemptions to the Freedom of Information Act, 5 U.S.C. § 552(b) (FOIA). Section 15 of the Federal Advisory Committee Act provides that the Academy shall make available to the public its final report unless the Academy determines that the report would disclose matters described in Section 552(b). In such case, the Academy "shall make public an abbreviated version of the report that does not disclose those matters." This abbreviated version of the report includes all of the committee's findings, recommendations, and other substantive material without disclosing materials described in 5 U.S.C. § 552(b).

Contents

Boxes, Figures, and Tables

BOXES

FIGURES

TABLES

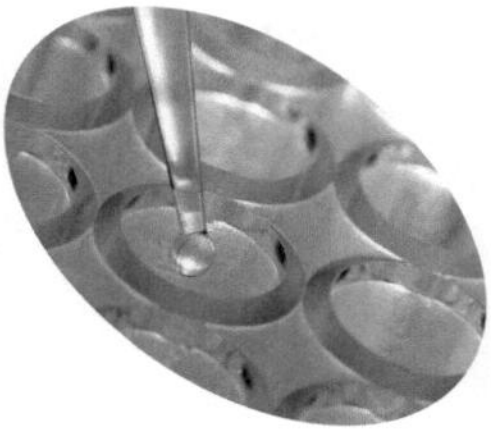

Acronyms and Abbreviations

APHL — Association of Public Health Laboratories

BAR — BioWatch Actionable Result
BDS — Post Office Biohazard Detection System
BTRA — Biological Terrorism Risk Assessment

CBRNE — chemical, biological, radiological, nuclear, and explosives
CDC — Centers for Disease Control and Prevention
CRP — Department of Defense Critical Reagents Program
CRS — Congressional Research Service
C_t — threshold cycle

DHS — Department of Homeland Security
DNA — deoxyribonucleic acid
DOD — Department of Defense
DOJ — Department of Justice
DTO — Defense Technology Objective

EPA — Environmental Protection Agency
EUA — emergency use authorization

FBI — Federal Bureau of Investigation
FDA — Food and Drug Administration
FSAPE — Federal Standards for Assay Performance and Equivalency

GAO	Government Accountability Office
HHS	Department of Health and Human Services
IOM	Institute of Medicine
JPEO-CBD	Joint Program Executive Office for Chemical and Biological Defense
LANL	Los Alamos National Laboratory
LLNL	Lawrence Livermore National Laboratory
LOD	limit of detection
LRN	Laboratory Response Network
MFI	median fluorescence intensity
MIQE	Minimum Information for Publication of Quantitative Real-Time PCR Experiments
MOU	memorandum of understanding
NAS	National Academy of Sciences
NPV	negative predictive value
NRC	National Research Council
NTTAA	National Technology Transfer and Advancement Act
OHA	DHS Office of Heath Affairs
OIE	World Organisation for Animal Health
OMB	Office of Management and Budget
PCR	polymerase chain reaction
PHAA	Public Health Actionable Assay
POD	probability of detection
PSAA	Public Safety Actionable Assay
PPV	positive predictive value
S&T	science and technology
SMPR	Standard Method Performance Requirement
SPADA	Stakeholder Panel on Agent Detection Assays
USAMRIID	U.S. Army Medical Research Institute of Infectious Diseases
USPS	U.S. Postal Service

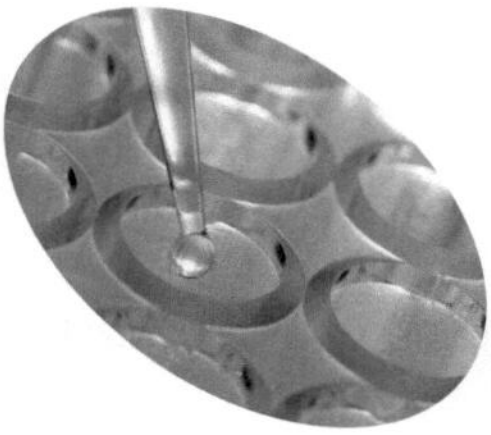

Executive Summary

BioWatch is an air monitoring system deployed in jurisdictions around the country with the goal of detecting the presence of certain high-risk pathogenic microorganisms. It relies on a network of federal and nonfederal collaborative relationships to be successful, and is one part of a larger array of disease surveillance, intelligence-gathering, and biomonitoring activities in support of public safety and health.

The assays used in the BioWatch system to detect the presence of pathogens in collected samples rely on the technique of polymerase chain reaction (PCR) to sensitively and specifically amplify target nucleic acid sequences. The program and its users need an understanding of each assay's performance characteristics in order to have confidence in the results and have the ability to appropriately interpret them. This confidence is partly provided through a performance standard, which gives information on the minimum requirements that must be met for the assay to be considered acceptable for its intended purpose and describes how testing to validate this performance is to be carried out.

The report discusses principles of performance standards, reviews information from several existing guidance documents and standards that might be applicable to BioWatch, and discusses assay testing efforts that have occurred or are ongoing. It provides recommendations on general principles and approaches for a performance standard and validation framework to meet BioWatch's mission. It also provides a starting point for an approach to *in silico* and laboratory assay characterization for consideration by BioWatch and its stakeholders, with the aim of providing a reasonable amount of information on performance parameters such as an assay's limit of detection, sensitivity, and specificity. As tasked, the report focuses on the PCR reaction portion of a PCR assay; although it discusses key processes before and after the PCR, these are not addressed in detail.

One fundamental aspect of assay validation for a system such as BioWatch should be recognized and clearly communicated to program stakeholders—BioWatch is designed to detect rare but high-consequence events and all assays will have some associated false positive and false negative rates. It is not practical in time, effort, or money to conduct laboratory-based assay performance testing using large enough numbers of replicates under enough different conditions to validate the assay's long-term false positive or false negative rate at the high levels of statistical confidence program users are likely to desire for operational decision making (e.g., to determine with statistical confidence that a false result will not occur more than once per *x* thousands or hundreds of thousands of samples). The assay design and laboratory performance validation that is the focus of the present report thus provides initial information that the assay is anticipated to perform reasonably well. It is critical for the system's users to understand the testing that has been undertaken and the limitations of the data, and to combine this knowledge with ongoing analysis of data obtained from verification in jurisdictional laboratories and from operational deployment of the assay in order to achieve confidence in the results and to facilitate responses to positive BioWatch detections. The report emphasizes the importance of communication and, particularly, of discussing the details of performance data with jurisdictional laboratory experts and officials. The report also emphasizes the importance of obtaining a better understanding of the microorganism background that is present in BioWatch jurisdictions, information that remains very limited.

Finally, the report considers how developments in technology, particularly in multiplex PCR and next-generation sequencing (NGS), can contribute to the ability of the BioWatch program to meet current and future challenges. Sequencing currently can be used to follow up on unexpected assay results from the jurisdictions. In the nearer term, targeted approaches coupled with NGS may be useful as a replacement for the current real-time PCR secondary assays because of their ability to analyze many more genomic regions for identification and characterization. The applicability of techniques such as metagenomic NGS to the program would be a longer-term prospect. The report recommends that the Department of Homeland Security and the BioWatch program continue to monitor and evaluate technologies as they develop. The program also should plan to work with laboratory users in the BioWatch jurisdictions, along with technology experts, to ensure that new technology brought into the program not only incorporates the best technology but also functions smoothly for the stakeholder community.

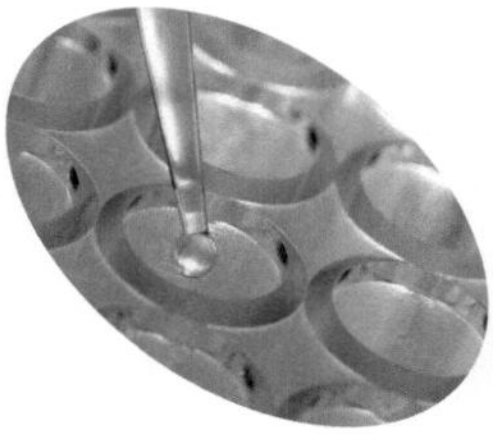

Summary

BioWatch is an environmental surveillance system designed to detect the presence of pathogenic microorganisms in air samples in order to provide early warning to local, state, and federal stakeholders in the event that one of these potential biothreat agents is present. BioWatch exists in the context of a number of complementary efforts to provide infectious disease surveillance (Figure S-1). These efforts include monitoring disease trends and case reporting for human and animal illnesses, syndromic surveillance (e.g., monitoring for notable increases in the incidence of symptoms such as fevers), and intelligence-gathering activities. The BioWatch system aims to complement these efforts in order to provide public health, emergency management, and law enforcement with an alert to the presence of one of the tested biothreat agents prior to the onset of large numbers of clinical symptoms. The system is thus meant to provide decision makers with the opportunity to initiate further investigations or undertake initial responses to minimize risks to public safety and health.

BIOWATCH MUST FUNCTION EFFECTIVELY AS A TOOL FOR DECISION MAKING

Although BioWatch is a federally funded air monitoring program under the Department of Homeland Security (DHS), the program's real impact lies in its ability to function as a decision-making tool for local and state jurisdictions that operate the program. The BioWatch system includes indoor and outdoor collectors deployed in jurisdictions around the coun-

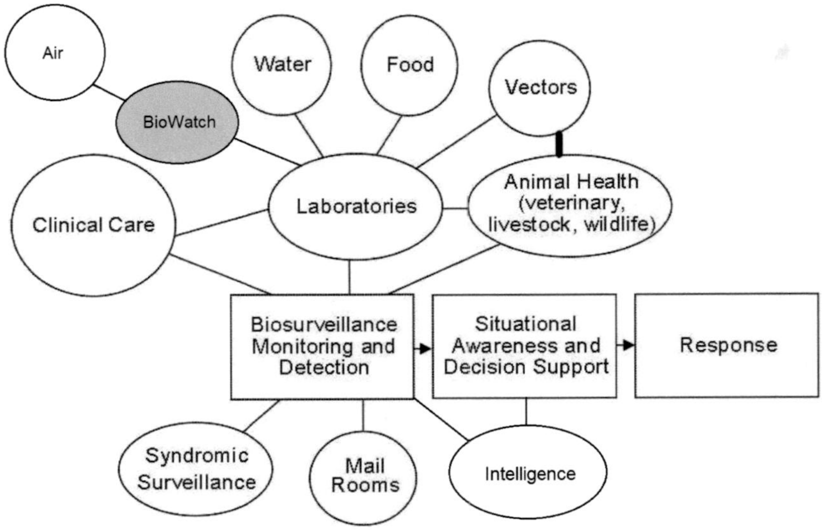

FIGURE S-1 Multiple sources of information provide infectious disease surveillance. BioWatch is an example of an air monitoring system and forms part of this network to support public health and security. Circles represent targets of surveillance information that inform monitoring and detection activities. The boxes show the process from biosurveillance through response.
SOURCE: Adapted from IOM and NRC (2011).

try, primarily in large urban settings, and relies on laboratory analyses largely conducted within the country's public health infrastructure. Each jurisdiction maintains a BioWatch Advisory Committee composed of law enforcement, public health, environmental health and protection, emergency management and medical services, National Guard Civil Support Teams, hazardous materials teams, and political officials that engages in close consultation with federal partners from a number of agencies. But ultimately, the actions taken or not taken based on results obtained from BioWatch aerosol collectors primarily impact the local level.

BioWatch relies on a series of process steps utilizing polymerase chain reaction (PCR) as its amplification and detection methodology to indicate whether or not nucleic acid from one of the tested pathogens is present on an air collection filter. Subsequent actions following a declaration of a BioWatch Actionable Result (BAR, indicating the presence of a pathogen's DNA) may be time-consuming, potentially disruptive, or expensive. Similarly, if the result of the BioWatch PCR assays is negative, individuals with responsibility for public health and safety need to understand the limits of the assays in determining that there is indeed no current threat to the public from one of the pathogens tested by the program.

It is critical that the BioWatch participants and the user community have confidence in the methodologies employed by BioWatch. An important element of this confidence is that the PCR assays used in the program have performance characteristics that have been validated and that provide proper and sufficient information to program stakeholders. This issue is the central topic of this report and is the focus of the committee's statement of task (Box S-1).

BOX S-1
Statement of Task

An ad hoc committee will be convened to conduct a study and prepare a report that will evaluate and provide guidance on appropriate standards for the validation and verification of polymerase chain reaction (PCR) tests and assays in order to ensure that adequate performance data are available to public health and other key decision makers with a sufficient confidence level to facilitate the public health response to a BioWatch Actionable Response (BAR).

Specifically the ad hoc committee will:

1. Determine PCR assay test and evaluation criteria that will provide a reasonable measure of confidence to federal, state, and local public health officials and key stakeholders.
2. Identify and evaluate the Stakeholder Panel on Agent Detection Assays (SPADA), the Public Health Actionable Assays (PHAA), and any other existing and proposed standards applicable for use in defining the performance (validation and verification) of PCR assays for the BioWatch and other programs to ensure confidence as identified in Subtask 1, above. Standards are to be evaluated in terms of performance, cost, and public health applicability.
3. Examine current PCR protocols used by the BioWatch program and other relevant biosurveillance programs and determine if the processes used to assess the performance of these protocols and assays are adequate to meet the standards identified in Subtask 2, above.
4. Determine whether improvements could be made by adopting changes based on the evaluation in Subtasks 2 and 3, above.
5. Determine if any existing standards approach is conducive, taking into consideration cost, schedule, and data requirements, to measuring performance of a PCR assay in multiplexed format.
6. In the event that no approach currently in existence is judged to be appropriate in Subtask 5, above, provide recommendations for aspects that a standard must include to measure performance of multiplex PCR technology.

PERFORMANCE STANDARDS AND VALIDATION SUPPORT CONFIDENCE IN RESULTS

The existence of appropriate performance standards and a rigorous process of validation provide the basis for confidence that PCR assay results meet users' needs. An assay performance standard describes minimum requirements that must be met for the assay to be considered acceptable for its intended purpose and describe how testing to validate this performance is to be carried out. To be applicable to BioWatch, a standard to characterize PCR assay performance and conduct assay validation would need to

- Establish that the assay has sufficient sensitivity to detect the release of a tested biothreat agent at a program-relevant concentration above the baseline of environmental background (i.e., it is fit for purpose);
- Establish that the assay has sufficient specificity to detect pathogenic strains of concern but not to cross-react with other strains and organisms, within acceptable program false positive and false negative rates;[1] and
- Determine that the assay is sufficiently robust for routine operational use by jurisdictional laboratories and is otherwise acceptable to program users.

From a practical standpoint, a decision maker would be most interested in knowing that the BioWatch system is sensitive enough to detect an amount of DNA reasonably expected to be present in a filter sample following a pathogen release of the scale for which BioWatch is designed. This question is complicated by the many factors involved in understanding overall system performance, including physical and biological characteristics of the threat agent particle, modeling analysis of environmental transport, the nature of the collection device and collection matrix, and others. The committee was not asked to examine the full spectrum of steps in the BioWatch system. Rather, the scope of the report is limited

[1]A false positive is a positive detection result for a given target when the target is not actually present. A false negative, on the other hand, is a negative result when the target is present. BioWatch has experienced prior positive assay results from detections of microorganisms in the environment that were not the result of intentional releases of a biothreat agent. Opinions vary on whether these results are "false positives" at the assay level, which may have performed as designed but detected naturally occuring microbes or revealed cross-reactivity with an organism having sequence similarity in the regions targeted by the PCR amplification reactions). A prior report on BioWatch classified such events as "BAR false positives" because they resulted in declaration of a BAR when intentional release had not occurred (IOM and NRC 2011). False negative rates are difficult to determine in routine or standard operation.

to the process for validating assay performance in a laboratory setting, commonly starting from extracted nucleic acid. Understanding assay performance even in this limited setting provides useful information in establishing the maximum sensitivity and specificity of the system, comparing assays when making decisions on which to consider deploying, and comparing the performance of different instrument platforms on a given assay. While a standard to characterize the analytical performance of an assay is thus an important component, program users and stakeholders need to be aware of its limitations in the full operational context.

The report considers a variety of parameters addressed by PCR performance standards and analyzes the implications of different types of performance approaches. Dimensions discussed in the report include

- The concentration of nucleic acid from a tested pathogen that will be detectable by the assay with a sufficiently high probability of detection (limit of detection [LOD]);
- The ability of the assay to successfully detect an identified minimal amount of nucleic acid from the set of organisms that makes up the targeted inclusivity panel (sensitivity);
- The ability of the assay to successfully not detect or cross-react with nucleic acid from the sets of organisms on the nontargeted exclusivity and environmental background panels (specificity);
- The ability of assay protocols to amplify and detect nucleic acid from filters that contain anticipated environmental background substances, such as dust and pollen, which can interfere with the PCR reaction (environmental interference);
- The ability of the assay to perform under real-world use, including robustness of the assay as it is performed by different operators and in different laboratories (robustness and reproducibility);
- The genetic diversity present in microbial species and strains, the continuing acquisition of new microbial genetic knowledge, and the need to regularly reassess assay performance (quality control and databases); and
- The controls, reference materials, and quality assurance processes associated with performance standards and validation (standards and quality assurance).

A BIOWATCH ASSAY STANDARD CAN BE INFORMED BY EXISITING GUIDANCE

The report reviews existing guidance on the process for conducting laboratory validation of the performance of PCR assays. A number of efforts have already been undertaken to provide a foundation of data on the performance of PCR assays that are used or could be used by BioWatch. Testing of relevant assays has been undertaken or is ongoing by groups such as the Los Alamos National Laboratory (LANL) and the Centers for Disease Control and Prevention (CDC), while previous efforts such as BioNet compared the performance of detection assays then in use by the Department of Defense (DOD) and CDC. The BioWatch program and its partners should be recognized for the valuable efforts they have already undertaken, as well as for their efforts to increase communication among the many users who participate in BioWatch. In concert with third-party contractors, DHS and the BioWatch program have designed laboratory arrangements for undertaking assay and device testing, assembled and quality-controlled DNA from a number of applicable inclusivity and exclusivity panel organisms, conducted testing to obtain system operational data, and established a quality assurance program with the jurisdictional laboratories in the network. These approaches represent appropriate options for characterizing real-time PCR assays for detection of biothreat agents.

The report seeks to build on this base of information. It discusses principles addressed by PCR performance standards, explores how these principles could be applied to meet the needs of the BioWatch program, and draws comparisons between these needs and existing PCR performance approaches, such as the Standard Method Performance Requirements (SMPRs) developed by the Stakeholder Panel on Agent Detection Assays (SPADA) and assay performance standards captured by the Public Health Actionable Assays (PHAA) program and Federal Standards for Assay Performance and Equivalency (FSAPE).

The available documentation on these standards provides a wealth of detail on the characterization and validation processes they recommend, including the testing that must be conducted and the controls to be included. SPADA generally addresses the needs of first-responder and private-sector detection development communities in support of public safety actions, while the PHAA program addresses public health decision making, including deployment of medical countermeasures. FSAPE was developed to support assay performance comparisons across federally owned and operated surveillance programs. These standards approaches have many similarities but also significant differences. For example, the number of strains required for inclusivity, exclusivity, and environmental background panels under PHAA and FSAPE are generally more numerous than the strain panels for testing under SPADA. These differences have implications for the time and costs necessary to comply with these standards. The approaches of the different standards also differ in how they were developed. SPADA meeting information is posted online and the resulting performance requirements are published through the *Journal of AOAC International*. PHAA and FSAPE are focused on federally owned programs; as such, their requirements are not published in the open literature.

The BioWatch program is a hybrid entity. Its results are used to support both public safety and public health actions. Although it is federally owned and relies on important federal relationships, it cannot succeed without close engagement with nonfederal partners. This fact makes it particularly important that any assay performance standard and associated validation program used by BioWatch be as transparent as possible to the range of the program's users. BioWatch particularly relies on the expertise of public health laboratorians to conduct and interpret assay results within their jurisdictions, and these experts need to be aware of the available performance data and be engaged in discussions surrounding assay performance issues that arise.

BETTER UNDERSTANDING ENVIRONMENTAL BACKGROUND WOULD IMPROVE THE UNDERSTANDING OF ASSAY PERFORMANCE

The current BioWatch approach focuses on testing for the presence of a subset of known biothreat agents considered to be of particular priority. Assay performance is likewise defined by testing sets of known inclusivity and exclusivity organisms available in genomic databases and in various strain collections. A fundamental issue is that knowledge of the relatedness of strains used for assay testing to microorganisms actually present in BioWatch environments remains very limited. An assay can be tested against one hundred strains, but if these are not the most relevant ones it will mean very little in practice. A frustration among BioWatch jurisdictions is the occurrence of positive detections that appear not to indicate biothreat agent releases. Many pathogens of relevance to BioWatch can occur naturally in the environment—certain pathogens may predictably trigger detections at specific times of the year under particular conditions in some locations. Certain assays also may cross-react with an environmental organism that is not yet included on the strain panels against which the assays are tested. To help address these issues:

Recommendation 1. The Department of Homeland Security should undertake further characterization of environmental background organisms over time and at various locations. This sampling could build on the existing BioWatch network by using previously tested filters as samples and should also include judicious sampling of water and soil samples. Operational assay performance data in individual jurisdictions based on positive screening results might provide a source of invaluable data, as well.

A PERFORMANCE STANDARD FOR BIOWATCH SHOULD ENCOMPASS A FULLER RANGE OF STEPS THAN CURRENT GUIDANCE

In addition to PCR amplification, multiple steps influence PCR assay performance (Box S-2) and there are multiple ways in which the term "PCR assay" can be defined. In particular, steps in the overall process that feed into the PCR amplification are critically important to overall performance. Sample collection, extraction, and purification methods affect the type and quality of materials that are available as the input to the PCR. The Minimum Information for Publication of Quantitative Real-Time PCR Experiments guideline (Bustin et al. 2009), for example, considers information on the sample processing steps to be essential to a description of the assay and its results.

Guidance available through the SPADA SMPRs and in PHAA and FSAPE documents does not provide explicit advice on the sample extraction and preparation methodology to be used and one might also expect the most appropriate sample processing and preparation method to vary based on the specifics of a particular assay and its intended purpose. The BioWatch program makes protocols available to its jurisdictional laboratories in order to provide standardized methods for the daily operation of sample processing and DNA extraction from BioWatch filters. The program has indicated that it is considering opportunities to improve these processes, but that such efforts are currently decoupled from PCR assay test and validation needs that are the subject of the present study. As requested by DHS, the present report focuses on the core PCR reaction in the overall process from collection of a BioWatch air sample to a result, and the key processes before and after the PCR are not significantly addressed. However, the committee notes that evaluating performance of PCR amplification starting with purified strain nucleic acid represents a simplified and only partial view of a complex problem. The committee encourages the BioWatch program to consider the critical roles of the steps surrounding the PCR amplification and concludes that it would strengthen a BioWatch PCR assay performance standard to explicitly consider guidance on steps such as sample extraction and preparation as a standard is developed:

Recommendation 2. A PCR assay performance standard that meets program and user needs should include development and implementation of standards that encompass all of the individual steps in the assay from sample collection, extraction, amplification, detection, reagent quality control, and data analysis. There should be an integrated assay standard, as well, that includes all of the steps of a complete PCR assay as a unit.

BOX S-2
A PCR Assay Comprises Multiple Steps

PCR is a widely used nucleic acid amplification technique, and its inventor, Kary Mullis of Cetus Corporation (Emeryville, CA), was awarded the Nobel Prize in Chemistry in 1993. PCR involves polymerization of nucleotides in a reaction performed and mediated by enzymes and thermal conditions, resulting in a theoretical doubling of the amount of targeted nucleic acid in each cycle. In the BioWatch program, PCR is the basis of the targeted amplification of specific nucleic acid sequences in order to identify whether nucleic acids from biothreat agents tested by the program are present on collected filters.

The report focuses primarily on the PCR amplification in the overall PCR assay, including testing undertaken to determine the ability to detect or not detect relevant pathogen strains and environmental organisms, and to determine the limit of detection.

A complete PCR assay includes steps in addition to the amplification reaction, including input sample extraction and preparation, reaction conditions, method of detection and instrument platform, and data analysis. Setting standards for PCR performance in a real-world setting cannot be done in isolation from these other dimensions of the assay, which affect overall performance. Although the report mentions aspects of the steps that compose the full assay, it does not, for example, make recommendations on specific sample processing methods, detection methodology, or result algorithms that should be used by the BioWatch program.

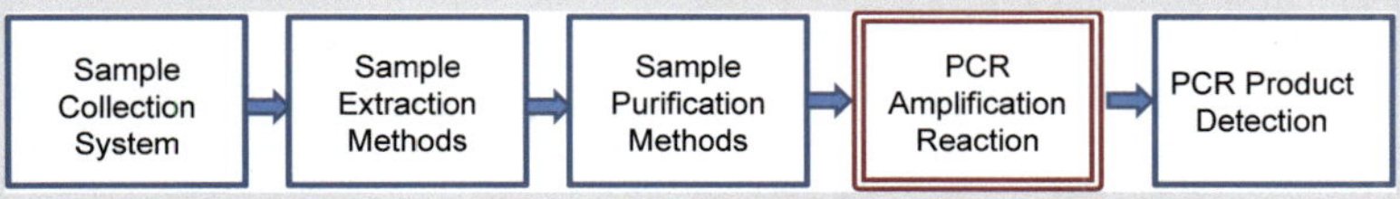

BOX FIGURE. Several key steps precede the PCR amplification. When combined with a detection mode, the set of steps forms a complete PCR assay. The report focuses primarily on PCR amplification and subsequent steps.

STARTING POINT FOR A BIOWATCH ASSAY PERFORMANCE STANDARD

The approach currently used by BioWatch—sensitive real-time PCR to amplify target regions of DNA from a set of known pathogenic organisms judged to be of highest risk—represents the best system BioWatch has been able to deploy given current technology and assay options. The program needs a basis for understanding the performance of its assays, as tested in laboratory settings, in order to inform a more comprehensive understanding of its system. The report highlights an example of a performance standards approach for conducting laboratory testing of assays that could provide reasonable statistical confidence for BioWatch users while seeking to keep numbers of tests and numbers of strains manageable for intended use. The specific details of the process ultimately used by the BioWatch program should be determined by the program and established in consultation with its stakeholder community.

In general, primary performance characterization and validation of an assay should be undertaken by the developer (or by a laboratory contracted by the developer). Further independent validation should be undertaken by DHS. Verification of assay and system performance should be undertaken by user laboratories independent from the developers. A smaller number of replicates can be used to verify assay performance (e.g., LOD, inclusivity, exclusivity, and environmental panel results) than was undertaken during full validation testing. The committee emphasizes that modeling and testing to understand performance needs in real operational contexts, monitoring sample flow through all stages of the system, and analyzing assay performance data in operational use must be conducted in addition to the initial laboratory assay validation. The committee recognizes that the BioWatch program has been undertaking and continues to undertake various types of system testing. In developing a performance standards approach to meet BioWatch's mission:

Recommendation 3. BioWatch should strive to test its PCR assays against as many known strains of its target agents and relevant near neighbors as possible. Undertaking all such testing through laboratory analysis alone is likely to be cost, time, and feasibility prohibitive. Increased *in silico* testing should thus be incorporated into the program to support targeted laboratory testing.

Recommendation 4. To support the goal of Recommendation 3, the BioWatch program should work with other federal agencies and international partners to ensure that as many strains as possible are sequenced to at least a high-quality draft level and the data and associated reference materials made available to PCR assay and device developers through reference databases.

Recommendation 5. The panels of inclusivity and exclusivity strains specified for laboratory testing should seek to balance broad phylogenetic and geographic diversity within the realm of practical testing. Inclusivity panel strains for laboratory testing should sample the genetic diversity represented by available strain collections with an emphasis on prioritizing those strains that cause significant morbidity and mortality, have high transmissibility, have a wide host range, or may be readily accessed by nefarious actors.

Recommendation 6. For laboratory testing of exclusivity panels (near neighbors and environmental organisms), a strategy of sample pooling is appropriate to increase testing efficiency. If a positive result is detected, each agent in the pool should be tested separately.

Recommendation 7. Assay performance in the presence of known environmental contamination (e.g., rail dust in a subway system) should be addressed by assessing the assay detection limit for a limited number of inclusivity and exclusivity panel agents spiked on jurisdictional background filters.

Recommendation 8. Federal agencies, including the Department of Homeland Security, the Department of Defense, the Department of Energy, the Department of Health and Human Services, and other partners, should collaborate to produce a reference document or database identifying which repositories contain strains or strain nucleic acids relevant for testing pathogen detection assays. Information on the requirements or procedures to obtain and use strains or strain materials (e.g., extracted nucleic acids) held by these repositories should be included, where possible. This information would provide a coherent source of information to both federal agency programs and to individuals or companies involved in assay and device development and validation.

Recommendation 9. The BioWatch program should institute a process to regularly review and update all assays in light of new genetic knowledge. The program should re-evaluate its assays through at least annual *in silico* screening to take advantage of new genomic information in databases, update signatures, and identify potential false positive or false negative issues. Similarly, the composition of inclusivity, exclusivity, and environmental panels used for testing should be reviewed regularly and laboratory re-testing conducted as needed. Annual *in silico* and periodic laboratory re-testing of assay performance should be part of any assay performance standard accepted for BioWatch.

The above recommendations provide strategies for a BioWatch PCR standard. Without duplicating the full level of detail contained in guidance documents such as the AOAC International validation guidelines, SPADA SMPRs, PHAA and FSAPE documents, and other sources, the committee also attempted to provide a specific starting point for methods to determine assay sensitivity and specificity that could be considered by BioWatch in developing a standard to meet program needs. Further details, including the statistical confidence and limitations associated with this approach, are discussed in Chapters 3 and 4. In addition, ***the committee concluded that ongoing analysis of assay performance data from verification in user laboratories and from operational deployment of the assay in the program is critical. It will not be possible to undertake laboratory validation testing using large enough numbers of replicates on samples from the full spectrum of potential inclusivity, exclusivity, and environmental organisms in the presence of a wide range of different types of potentially interfering substances and different filter conditions to characterize the long-term false positive and false negative rates of an assay to a level of statistical confidence users may desire operationally.*** Nevertheless, some reasonable compromise must be achieved in order to undertake performance characterization and validation work on assays deployed or being considered for deployment in the program. Understanding and clearly communicating the strengths and limitations of the laboratory data obtained from the PCR assays is crucial to interpreting results and using them as a basis for taking action.

Recommendation 10. The following approach can serve as a starting point for a standard to provide confidence in PCR performance while seeking to undertake a reasonable amount of laboratory characterization and validation testing:

- A reasonable approach to determining an assay's analytical LOD in a laboratory setting is to conduct serial dilution at a range of concentrations bracketing the estimated LOD, using $n = 60$ replicates with acceptance criteria of at least 58/60 for a given concentration, followed by appropriate curve fitting. LOD testing should also be conducted in the presence of realistic background matrix, such as previously tested BioWatch filters, to gain an understanding of how the LOD may be affected by operational background.
- The use of *in silico* screening presents a relatively fast and low-cost way to predict one part of assay performance using as many strains as are available as high-quality sequences in reference databases. *In silico* screening should be undertaken using the set of PHAA panels, if these strains' sequences are available.
- The strain panels recommended by SPADA represent a good starting point for laboratory validation testing of BioWatch assays and were developed with stakeholder participation. However, many were developed in 2010-2011 and should be reviewed and updated to account for new genetic knowledge.

- The strategy used by LANL for testing detectability or nondetectability of inclusivity and exclusivity strains represents a reasonable model for BioWatch (testing n = 20 replicates, followed by testing 20 further replicates if an unexpected detect or nondetect result is obtained, with acceptance criteria of 20/20 or 39/40).

It is not possible to predict the precise costs of the committee's proposed validation framework. Data available to the committee on costs of assembly and quality control of relevant strain materials and to undertake validation testing following SPADA and PHAA guidance are available in Table 3-3 for comparison. The costs of the committee's approach appear to be comparable in scope.

It should also be noted that the BioWatch program currently employs an initial screening assay, provided through the DOD's Critical Reagents Program, to detect a single target nucleic acid sequence per tested pathogen. If this screening assay yields a positive detection, a secondary assay panel, provided through the CDC Laboratory Response Network, is used to test for three or more additional pathogen sequences. A basic performance standard and validation framework can be appropriate for assays used for both screening and for secondary testing. However, the assessment of testing results in order to determine whether the assay meets program needs will be different depending on the way(s) the assay is to be used in the program. Because of the importance of specificity to the performance of a secondary assay, the program might conclude that it is useful, for example, to conduct laboratory validation using additional inclusivity and exclusivity strains than were used for an initial screening assay. An approach of serial testing or "tiers" of recommended testing makes sense from a practical perspective given limited resources. Thus, an assay meant to be used for secondary verification or confirmation might be validated against a larger number of inclusivity and exclusivity strains than an assay meant to be used as an initial screen.

ALTERNATIVES TO THE CURRENT SCREENING AND SECONDARY ASSAY APPROACH

The report also explores several options that BioWatch could consider as alternatives to its current model of initial screening and secondary PCR assays, which would have implications for the performance characterization and validation process required. These include the following:

- Develop subsets of assays for each of the tested BioWatch agents rather than a single set of screening and secondary assays (e.g., incorporate additional types of rule-in and rule-out assays for certain pathogens, along with decision algorithms on when to use which tests);

- Develop and deploy multiplex PCR assays; and
- Incorporate next-generation sequencing (NGS) into the program as technology and software development makes this capability increasingly feasible.

Multiplex PCR

Each PCR assay currently used by BioWatch amplifies a single target sequence of pathogen nucleic acid (e.g., the assays are run as singleplex reactions). Assays also can be designed to detect multiple target sequences per reaction and technology in this area is moving fast, particularly in the development and use of multiplex assays in clinical settings. Multiplex PCR assays could be used by BioWatch and have the potential to reduce the labor incurred by jurisdictional laboratories as they conduct routine filter testing.

Appropriate standards and validation guidance recently have emerged to address the additional issues that arise when characterizing and validating multiplex assays. Because PCR remains the core detection technology, many of the same performance principles apply. Additional types of testing will need to be conducted for multiplex assays, for example, to determine that each of the target sequences can be sufficiently amplified in the reaction and that none of the components in the reaction interfere with each other. Other potential issues involved in developing successful multiplex assays include determining the concentrations of different reagents, primers, and probes (including the need to limit certain reagents) when aiming to amplify and detect two or more pathogen targets, and the potential for nonspecific amplification products to arise. A further complication arises when making subsequent changes to a previously validated multiplex assay, in terms of the extent of performance re-testing that is required.

Recent guidance from the Food and Drug Administration (FDA) provides a path forward in such circumstances, targeted to those submitting materials under FDA regulatory approval. For example, FDA guidance suggests that changes to already approved assays (such as inclusion of a new target) can be made using subsets of testing to confirm performance, rather than requiring the full range of validation testing to be repeated. ***The committee concluded that existing guidance on singleplex PCR performance for environmental biodetection assays in concert with FDA information should provide a good starting point for the performance testing and validation of multiplex assays by the BioWatch program.***

Additional considerations will arise with multiplex assays as compared to singleplex assays, due to the need to validate mixtures of targets and to investigate potential cross-reactivities. However, the development of multiplex assays for use in BioWatch appears reasonable and could likely be achieved without an unacceptable loss of sensitivity and specificity for program users. More important, however, the use of multiplex assays in BioWatch would require agreement from program stakeholders, particularly the directors of the jurisdictional laboratories who conduct assay analyses and lead public health officials with responsibility for using the information obtained. Obtaining this agreement will require improved mechanisms for sharing and discussing detailed assay performance data. ***The additional issues that arise when designing, validating, and deploying multiplex assays will need to be considered when making the decision to switch to a multiplex format.***

Next-Generation Sequencing

Moving forward, the committee urges the BioWatch program to consider the ways in which emerging technology, particularly next-generation sequencing, can help the program address some of the challenges inherent in its current approach, which is focused on detecting a subset of high-priority, previously known pathogens. Next-generation sequencing presents possibilities for the program to move to a system that can be significantly more flexible and broad-spectrum in identification. The program already can use sequencing as a tool to help resolve unusual assay results. In this way, sequencing can function as a tertiary confirmation and characterization tool. Several additional ways that NGS could be used by the program and current barriers to implementation are summarized in Table S-1.

In the relatively near term, targeted NGS approaches may become useful to the program in place of current secondary PCR assays because of the ability to analyze more regions for identification and characterization, thereby improving specificity of detection. Metagenomic sequencing, on the other hand, could assist in characterizing environmental background and might eventually enable the implementation of universal assay protocols that would minimize the effort and costs needed to continually update and revalidate PCR primers, probes, and assay conditions. Sequencing and associated bioinformatics are not yet at a stage at which they could replace PCR assays as the BioWatch program's primary detection methodology because of cost, turnaround time, lack of standardized reference databases for comparison, and complexity of data analysis. Incorporating sequencing-based approaches into BioWatch will also require jurisdictional training, expertise, and infrastructure that do not

TABLE S-1 Applications of Next-Generation Sequencing for BioWatch

	Related Terms	Purpose	Advantages	Key Challenges / Barriers to Implementation			
				Turnaround Time	Cost	Complexity of Bioinformatics Analysis	Challenges of Interpretation
Metagenomic sequencing	Shotgun sequencing, random sequencing, unbiased sequencing	Primary BioWatch assay	Most inclusive	X	X	X	X
		Environmental and detection filter background measurement	Enable detection of "natural" positive controls (fungal spores, etc.)		X	X	X
			Better understand sources of false-positive signals				
			Improve design of primers and probe targets used for primary screening				
		Novel pathogen discovery	Identification of novel or divergent pathogens that would go undetected by existing tests		X	X	X
Microbial whole-genome sequencing	Whole-genome resequencing, *de novo* genome assembly	Optimize design of primers and probes	Will identify new targets for PCR and improve primer / probe design		X		
		Expand reference databases	Will better capture target diversity and inform primer / probe design		X	X	
Targeted sequencing	Amplicon sequencing, biased sequencing, conserved region sequencing	Investigate false-positive results	Better understand sources of false-positive signals	X	X		X
		Primary BioWatch assay	More informative than "yes-no" result from real-time PCR; more inclusive of diverse strains	X	X		

currently exist in all laboratories. Nevertheless, exploring these options and incorporating them into the program through pilot testing will help BioWatch meet future challenges, such as the potential emergence of new or unknown pathogens or the possibilities for nefarious activities such as bioterrorism enabled by synthetic biology.

Recommendation 11. The Department of Homeland Security should monitor developments in technology, such as next-generation sequencing, as speed increases, cost decreases, and required informatics and databases improve. A program to evaluate technology and incorporate it into BioWatch would enable the program to improve its ability to address current challenges and enable it to meet those of the future. However, the program should plan to work with at least some early adopter laboratory users in the BioWatch jurisdictions, along with technology experts, to make sure that new technology brought into the program not only incorporates the best technology but also functions smoothly for the stakeholder community.

THE IMPORTANCE OF COMMUNICATION TO PROGRAM SUCCESS

Finally, one of the primary messages of the report is that the network of BioWatch stakeholders should cooperate and be integrally involved in the process of establishing standards that will meet the community's needs and should have access to the relevant data to help them understand the performance and limitations of the assays in order to enable appropriate interpretation of the results. Building additional, specific mechanisms into the program to share assay and system performance results, discuss what these performance results reveal about the limits of the assay and the translation of assay results into actions, and share applicable detailed assay performance information across agencies such as DHS and CDC and with appropriate individuals within the network of jurisdictions appears to the committee to be a critical need. Even while recognizing the necessary security constraints, this goal should be achievable. Addressing this issue would go far in ensuring confidence in the system and its results in order to facilitate responses to a BAR.

Recommendation 12. A systematic process should be instituted to identify the root cause of positive assay results that do not appear to represent the targeted microorganism and to determine appropriate corrective actions to address the cause(s). This feedback and revision process should include the participation of laboratory experts in the jurisdictions and federal experts.

Recommendation 13. Communication and information-sharing are necessary to establish confidence in a system. The BioWatch program, relevant federal agencies, and local and state jurisdictions should expand the communication and data-sharing that occurs among the network of federal and nonfederal partners involving both the screening and the secondary PCR assays. This effort would assist in establishing acceptable performance standards, enhance informed data interpretation and decision making, improve the ability to undertake root cause analysis of assay issues encountered by jurisdictions, and enable the collective identification and dissemination of actions as part of robust quality assurance.

The ability to detect pathogen nucleic acid from BioWatch collectors in such a way that the information is usable for decision making and is obtained on a sufficiently rapid timescale and at reasonable cost is fundamental to the acceptability and success of the program. The report aims to contribute to the discussions surrounding performance standards and validation approaches that can meet these BioWatch program needs. A valuable repository of data and performance guidance can be drawn on to support BioWatch's mission and the program has a strong foundation from which it can develop performance and validation standards. Although it will require the participation of multiple groups, there is an opportunity to further strengthen the system by incorporating additional data-sharing and performance review features into the existing validation and quality assurance framework.

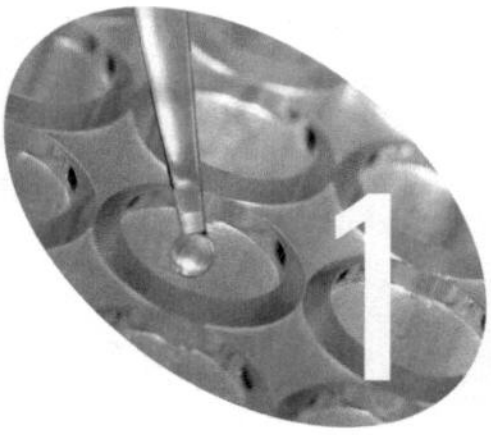

1 Introduction

The United States has a strategic interest in identifying and responding effectively to potential disease outbreaks, whether they result from the spread of naturally occurring pathogens or are due to accidental or intentional release of infectious microorganisms. The web of systems needed to achieve this goal range from environmental monitoring to clinical and public health epidemiology to intelligence community assessments of bioterror threats. Implementing methods to detect the presence of pathogens of concern in the environment forms one component of the strategy. As defined by Homeland Security Presidential Directive 21, biosurveillance is

> the process of active data-gathering with appropriate analysis and interpretation of biosphere data that might relate to disease activity and threats to human or animal health—whether infectious, toxic, metabolic, or otherwise, and regardless of intentional or natural origin—in order to achieve early warning of health threats, early detection of health events, and overall situational awareness of disease activity. (White House 2007)

The biosurveillance umbrella captures multiple activities of threat awareness, detection, characterization, and information sharing in clinical and environmental settings. The *National Biosurveillance Science and Technology Roadmap* (NSTC 2013), for example, describes over 20 programs, networks, and activities from across the government that contribute to the National Strategy's goals. Table 1-1 lists selected documents that address strategic priorities for biosurveillance activities. BioWatch, the focus of this report, is one component of the overall system.

TABLE 1-1 Selected Federal Strategy Documents Related to Environmental Biosurveillance

Year	Document	Description
2004	Homeland Security Presidential Directive 10: Biodefense for the 21st Century	Describes pillars of national biodefense preparedness including threat awareness, prevention and protection, surveillance and detection, and response and recovery (White House 2004)
2007	Homeland Security Presidential Directive 21	Establishes a strategy for public health and medical preparedness (White House 2007)
2009	National Strategy for Countering Biological Threats	Provides a planning framework and describes seven objectives to reduce threats posed by biological risks (NSC 2009)
2011	A National Strategy for CBRNE Standards	Articulates goals to enhance coordination on standards and to establish conformity assessment capabilities (NSTC 2011)
2012	National Strategy for Biosurveillance	Promotes strengthened biosurveillance capabilities (White House 2012)
2013	National Biosurveillance Science and Technology Roadmap	Identifies research and development needs to enhance the effectiveness of biosurveillance activities (NSTC 2013)

BIOWATCH DETECTS AIRBORNE RELEASES OF SPECIFIC BIOLOGICAL AGENTS

The BioWatch program is part of a layered approach to rapidly detect and respond to exposure to airborne biological agents of concern. A brief overview of the program is presented below and additional information is provided throughout the report. Certain details that the committee used to inform its analysis have been determined by the government to be exempt from public release under 5 U.S.C. § 552(b)(7); further information on these aspects is provided in Appendix A.

BioWatch is an environmental monitoring system developed to detect and characterize genomic material from a selected set of aerosolizable pathogens.[1] The program is operated by the Department of Homeland

[1] The decision on which pathogens are included for detection in the program is informed by risk assessments such as the Biological Terrorism Risk Assessment (BTRA) and material threat assessments for specific agents. These risk assessments are developed by the DHS Science and Technology Directorate as part of DHS's mandate under several Homeland Security Presidential Directives. The BTRA has generally been undertaken at 2-year intervals, and DHS reportedly reviews the alignment of BioWatch with the BTRA when a new assessment is released (GAO 2012b).

Security (DHS) and managed through its Office of Health Affairs. The DHS Science and Technology Directorate also supports program goals by undertaking activities in research and development that contribute to its operations. BioWatch was first deployed in 2003 in outdoor urban settings, amid concerns about potential damage that would be caused by a bioterrorism attack that could affect significant numbers of people.[2] The current system (sometimes referred to as the "Generation 2" or "Gen-2" system) includes collectors in over 30 jurisdictions and makes use of both outdoor and indoor collectors; systems also are deployed during selected special events. By 2013, reportedly more than 7 million polymerase chain reaction (PCR) assays had been conducted during routine daily operations to monitor for genetic material from the set of microbial agents for which BioWatch tests (IOM and NRC 2014).[3]

BioWatch comprises a network of components, processes, and participants that interact to capture genetic information from aerosol samples and determine whether and what actions should be taken based on the results. The context of BioWatch operations and the complexities and limitations of the role it plays should be appreciated. To detect an intentional release of biological material from air samples involves a host of factors affecting overall system performance. Environmental conditions, such as wind speed and direction, and particle parameters, such as size and density, will alter the transport and dissemination of biological aerosols. The location of collection devices and the spacing between collectors relative to the location of a released plume will affect signal strength at the collection device and whether more than one device will register a detection. The method of collection, particle size cutoff, and air sampling volume will affect the amount of biological material that is present in a sample. Finally, characteristics of the pathogen itself, such as the infectious dose and whether or not the agent is viable, will influence whether exposure to a particular quantity of an airborne microorganism is likely to produce disease.

From a practical standpoint, decision makers will want to know whether the BioWatch system can reliably detect airborne biological material at a level that would indicate the occurrence of a release of the size and scope for which BioWatch was designed. This question can be roughly divided into two parts: (1) what amount of material from a tested

---

[2] BioWatch's mission supports goals articulated in the 2003 State of the Union Address of an "early warning network of sensors to detect biological attack" (Bush 2003). BioWatch's mission is to "Provide, maintain and support a continuous aerosol bio-terrorism monitoring capability in selected metropolitan areas" (Walter 2014a).

[3] The locations of BioWatch air collectors and certain operational details of the program are not considered public information. See Appendix A for additional program and operational details.

agent needs to be deposited at the collection device to indicate with high probability that a release has occurred (or alternatively, what probability of a release is indicated by the presence of x amount of biological material at the collection device), and (2) what is the probability that the assays used in testing the sample will be able to reliably detect this level (or a lower one) of material.

The committee was not asked to examine and does not have information on the first of these components—the minimum size of a release event that BioWatch should be able to detect and how this might translate to a predicted minimum amount of material at a collector. Rather, the scope of this report is limited to the second part: How can the assays that test for the presence of a biological agent be validated so as to give the BioWatch program and its stakeholders confidence in their performance? This piece is in many ways the most straightforward. As a result, it is also worth noting that high confidence in the PCR assay is only one dimension of the system, and should not give users an overly optimistic sense of the performance of the system as a whole without further system testing.

The main processes in BioWatch are shown in Figure 1-1, which depicts steps from filter collection to assay interpretation and decision making.

BioWatch collectors continuously draw air through a dry filter where particulates are deposited. These filters are collected manually at regular intervals, usually once a day, and taken to participating diagnostic laboratories for analysis. The laboratories that analyze the filters collected from BioWatch air samplers are generally co-located with state and local public health laboratories, which maintain a dedicated BioWatch capacity. Most of these public health laboratories are members of the Centers for Disease Control and Prevention's (CDC's) Laboratory Response Network (LRN), which forms part of the national infrastructure to respond to public health emergencies and to potential bioterrorism incidents.

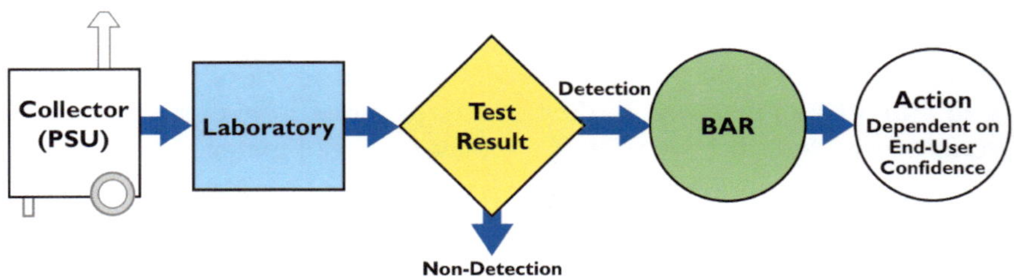

FIGURE 1-1 Schematic of the steps in the BioWatch process from air collection to assay interpretation and decision making. PSU = portable sampling unit; BAR = BioWatch Actionable Result.
SOURCE: Walter (2014a).

At the laboratories, DNA is extracted from the air filters and targeted regions from the genomes of the pathogens of interest are amplified and detected using the method of real-time PCR (see Box 1-1). For these analyses, a set of PCR primers specific to DNA from the microorganisms of BioWatch concern is used for the PCR assay. For each tested pathogen, the laboratories use an initial screen targeting a single nucleic acid target sequence (a signature) for preliminary identification of candidate microorganisms. If this initial screen yields a positive result, a secondary panel of assays directed against three or more additional nucleic acid signatures is used to verify the presence of the agent.[4] The BioWatch screening and secondary PCR assays currently are used in an integrated fashion by laboratories in the program (e.g., a BioWatch Actionable Result [BAR] generally is not called only on the basis of the results of the screening assay).

Possible Outcomes Arising from the PCR Assay

Based on the readouts from the PCR assays and the associated assay decision algorithms, there are four possible outcomes:

- True positive: the assay indicates the presence of the tested agent's DNA, and the agent's DNA is indeed present in the sample.
- True negative: the assay does not detect the agent, and the tested agent is indeed not present in the sample.
- False positive: the assay indicates that the tested agent is present in the sample, but in fact, the agent is not in the sample.
- False negative: the assay indicates that the tested agent is not present in the sample, but in fact, the agent is in the sample.

No assay will ever be free of false positive and false negative results, and these results are dependent on how one defines these terms in practice. The potential for fundamental limitations of a test's design (tradeoffs between sensitivity and specificity, for example) and instrument and operator failures means that some likelihood of error will always be present. Understanding the nature, origins, and probability that a potential error may occur assist in the interpretation of results and provide a measure of confidence for the user.

---

[4] The report uses the term "secondary assay" throughout to refer to this step, recognizing that it actually entails the performance of several individual PCR assays to detect several distinctive genetic signatures for the pathogen. BioWatch program materials frequently refer to this as the "verification assay" or verification assay panel; the committee prefers the term "secondary" to distinguish the assay from the processes of assay validation and verification.

BOX 1-1
Real-Time PCR as a Pathogen Detection Method

PCR is a widely used nucleic acid amplification technique, and its inventor, Kary Mullis of Cetus Corporation (Emeryville, CA), was awarded the Nobel Prize in Chemistry in 1993. PCR involves the polymerization of nucleotides in a reaction performed and mediated by enzymes and thermal conditions, resulting in a theoretical doubling of the amount of nucleic acid in each cycle of PCR. Real-time PCR adds a detection strategy to the amplification process—a common method is through the use of fluorescence, as shown in the figures below. A series of temperature changes is used to separate strands of DNA, allow primers and a probe to hybridize to complementary sequences in the target region for amplification, and enable polymerase to synthesize new nucleic acid in this region. As the polymerase synthesizes the new nucleic acid strand, it degrades the probe, releasing a fluorescent reporter whose emission can be detected. The greater the fluorescent signal, the greater is the amount of target in a sample. Other technologies can use a DNA-binding dye instead of a fluorogenic probe to measure DNA amplification. The fluorescent signal can be monitored at each cycle to create curves such as those below, with associated threshold cycle (C_t) values for when the signal crosses the background threshold. Alternatively, the strategy of end point detection, measuring only the final signal rather than the intermediate amplification curves, may be used. In this case, results are frequently reported as median fluorescence intensity values.

PCR methods can be used to detect nucleic acid from agents of interest such as those tested by BioWatch. Using a pathogen's genetic sequence and genetic information from related microorganisms (such as closely related but nonpathogenic species to reduce cross-reactivity), specific sets of primers and probes are designed. The goal is to amplify and detect a DNA target (an amplicon) that is distinctive for the tested organism. The curve of the detected fluorescence provides information on whether the DNA signature of the pathogen is present in the sample as well as information on the starting amount of target DNA. For BioWatch, there are multiple steps from a pathogen in the air to a sample of extracted DNA used in a PCR, and so the relative amount of DNA in the PCR sample can be informative but does not directly provide the original pathogen concentration.

During a PCR assay, a false positive detection could result from a number of factors, such as cross-reactivity of assay primers and probes (both types of short oligonucleotides that hybridize to target DNA) or the presence of a closely related but as-yet unsequenced organism occurring naturally in the environment. As was emphasized to the committee dur-

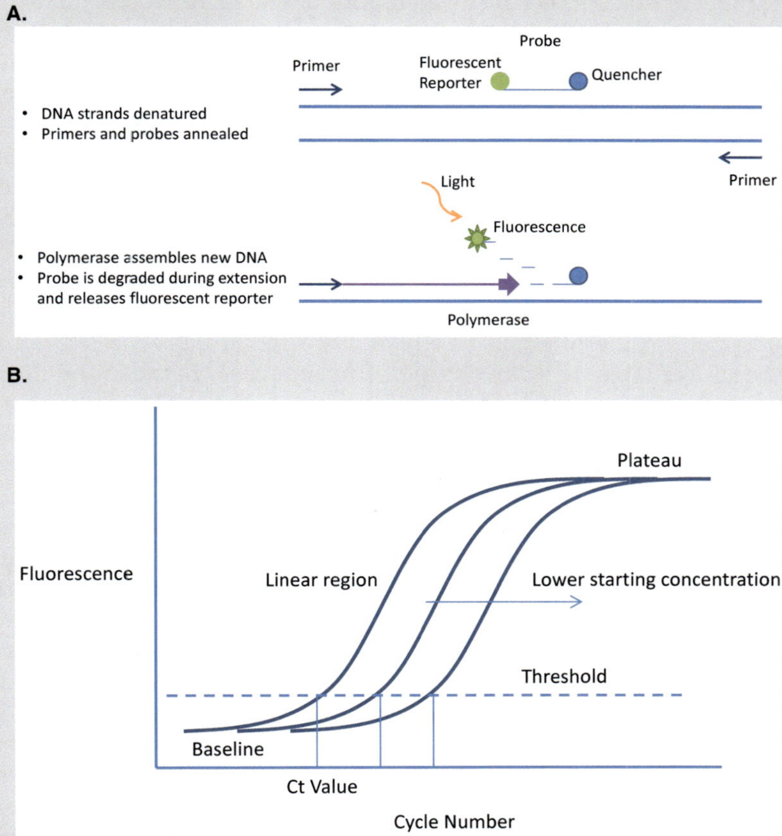

BOX FIGURE. Depiction of a method of real-time PCR that makes use of a fluorescent reporter molecule. **A.** The fluorescence of the probe is normally quenched. During amplification, the probe is degraded, and emissions from the fluorescent reporter can be detected. **B.** If the sample contains the signature DNA for amplification, the fluorescent signal will increase with each cycle. A threshold when signal crosses background is set, and the cycle number at which this threshold is reached is reported as the C_t value.

ing its data-gathering and has been highlighted in other reports, the public health system is called upon to do many things with limited resources. Responding to a detection that turns out to be a false positive places considerable drain on jurisdictional resources and could undermine confidence in the system when a true positive occurs. Over the history of the program, PCR assay positive results that appear not to have been true biothreat agent detections have been reported (Garza 2012; U.S. House of Representatives 2012a), resulting in changes to certain assays and decision algorithms. In the current BioWatch system, a sample yielding a positive result in the screening assay undergoes a secondary assay for the presence of several additional target sequences. A BAR is not normally called until after this secondary step, reducing the likelihood that jurisdictions and federal partners will mount a full-scale response that turns out to be unnecessary.

On the other hand, a false negative result could be catastrophic if a pathogen release is missed. At the collector, factors such as air temperature, relative humidity, and wind may influence how airborne particles are deposited on the filter, all of which can affect the type and amount of material available as a starting point for analysis. Because the filters capture all particulates of a given size and not only genetic material, the mixture of material can be complex and may contain molecules that present challenges for nucleic acid extraction, may cause loss and degradation of the sample, or are capable of inhibiting PCR amplification. Methods used for extracting DNA from the filter and preparing the sample for downstream analysis likewise affect the input material for the PCR assay. In the PCR assay itself, the presence of molecules that inhibit PCR, operational errors, instrument failures, or other errors in PCR amplification and detection could all produce a negative result despite the actual presence of pathogen DNA. Additionally, the concentration of the pathogen on the filter may be below the threshold of detection of the given assay. To help identify errors resulting from instrument failures or contamination, samples are run with sets of positive and negative controls. However, since specific sequences are targeted by the primers and probe, a failure to detect also could be the result of mutations in the pathogen's targeted areas, loss of targeted sequences (e.g., in assays that target a bacterial plasmid), or the result of a previously unknown or manipulated strain of the pathogen. Addressing these concerns poses a greater challenge and is discussed as part of PCR assay performance validation in Chapter 2.

A point is sometimes raised that the PCR assays used by BioWatch do not provide information on agent viability, and thus positive detection results do not necessarily indicate the presence of an active risk to public safety and health. Viability information is extremely useful in making further decisions on what public health actions may be necessary. On the other hand, the committee noted that detecting the presence of DNA from even nonviable agents on a BioWatch filter is a situation that law enforcement, public health, and others need to be aware of and that should be further investigated. It could, for example, indicate the presence of a failed release. Thus, the committee sees an important role for the BioWatch assays separate from the issue of additional testing to characterize whether the agents are viable. The committee understands that the BioWatch program may use RNA assays in viability testing as part of additional investigations following a BAR. RNA testing also can be used to detect the presence of DNA-based organisms, including bacteria or parasites (Murphy et al. 2012). The sample collection method would need to be one that preserves the presence of RNA, which is more labile than DNA. The report does not delve further into this strategy.

It is also important to emphasize that BioWatch is designed to detect very rare events, but ones that would be of high consequence should they occur. The fact that any assay will have some associated false positive and false negative rate, combined with the very low probability of a biological agent release, implies that most BARs will result from detections that are not actual releases or attacks. This type of event has been observed over BioWatch's history and is inherent in the nature of this type of system. As a result, it is critical to be able to analyze a BAR in the context of other sources of information in order to inform decision making.

BioWatch needs to serve public safety and health as a tool for high-visibility decision making. Because there are serious consequences from both false positive and false negative results being passed forward, the program has defined acceptable limits for these rates as part of testing within its quality assurance program (which includes the set of steps from DNA extraction from a filter sample through to the interpretation of a result following PCR amplification and detection; see Appendix A for these limits and a discussion of what they encompass). However, the overall system performance will not be better than the performance of any of the individual components. Because real-time PCR is the core detection method currently used by BioWatch, assay performance characteristics, including false positive and false negative rates, provide valuable information to the program's managers and users.

BioWatch Actionable Result

The responsibility for declaring a BAR following the positive results of the PCR assays rests with the director of the local laboratory that conducted the sample analysis, or an appropriate designee, and technical consultation with partners in DHS and CDC can occur before making the determination. As described to the committee, a BAR indicates that nucleic acid considered distinctive for a tested organism is present on a collected filter and can provide qualitative indications of the concentration of material along with the location and time of filter pickup. The BioWatch program considers a BAR to be a detection that activates notification protocols and warrants some form of response from the affected jurisdiction and the federal government. The response usually involves a local conference call to discuss the BAR and begin to obtain information on the local response strategy, and subsequent national conference call between the jurisdiction and federal partners to share information and put the BAR into context. As was presented to the committee, a BAR "does not necessarily mean that a terrorist attack has occurred; a viable biological agent was released; the agent is infectious; [or] there is a risk to the public's health" (Walter 2014a). Post-detection situational assessment activities (e.g., further environmental sampling, analysis of additional threat-related information, or pathogen characterization) are frequently undertaken to provide data to better understand the circumstances and potential risk to public health to inform decision makers. Figure 1-2 shows the scope and approximate time line of BioWatch operations.

The BioWatch program relies on the expertise and collaboration of the public health community and other stakeholders for analytical laboratory capacity and, more importantly, in determining appropriate responses to a positive assay result. The network of BioWatch partnerships required

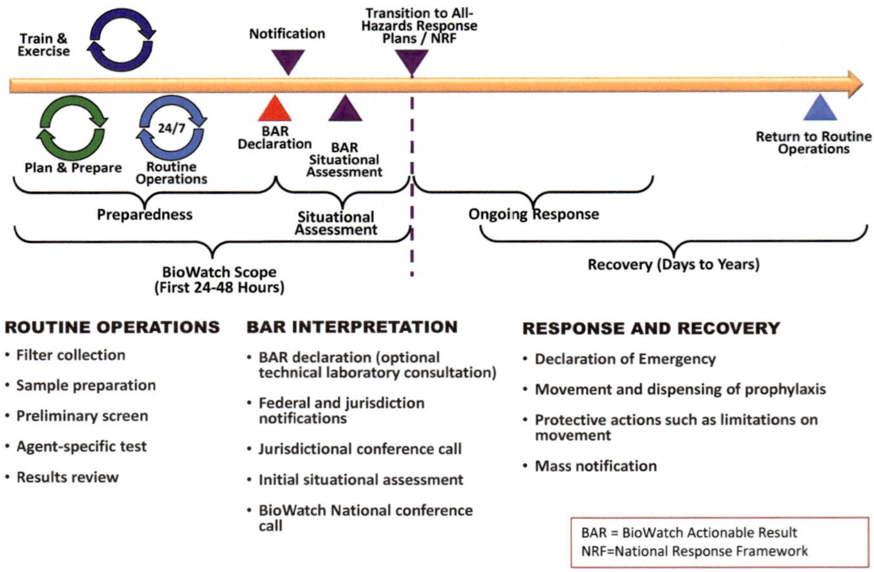

FIGURE 1-2 The BioWatch program maintains routine operations in support of preparedness goals and operates in the early notification and assessment stages of a potential event involving release of an aerosolized pathogen. Another important component of response and recovery is cleanup.
SOURCE: Walter (2014a).

for successful operation includes DHS, state and local officials from public health, law enforcement, emergency management in the BioWatch jurisdictions, and federal partners such as national laboratories, CDC, the Environmental Protection Agency (EPA), the Department of Defense (DOD), and the Federal Bureau of Investigation (FBI). The EPA is involved in additional environmental sampling activities after a BAR and any necessary decontamination, while the Department of Justice and FBI would play significant roles in any investigations of alleged pathogen release and attribution.

PCR ASSAY STANDARDS CONTRIBUTE TO PROGRAM SUCCESS

BioWatch's mission is to provide timely warning of the presence of a set of targeted pathogens of concern. Success in this mission will be a useful tool for decision makers. It is therefore critical that the BioWatch participants and the user community have trust in the results. An important element of this trust is that the PCR assays have performance characteristics that have been validated and that provide sufficient confidence to program stakeholders. This point is the central topic of this report.

An assay performance standard describes the minimal requirements that must be met for the assay to be considered acceptable and describes how testing to validate this performance is to be carried out. PCR performance standards include items such as the bacterial or viral targets, limits of detection (LOD) for a target, reproducibility of detection, dynamic range of detection, the strains that the assay will detect, including strains or closely related strains it should not detect, and associated confidence limits of the results. A useful BioWatch standard is required to be "executable in a reasonable time, affordable, [and to] maximize confidence in results" (Walter 2014a). A standard or set of standards also needs to be acceptable and transparent to the program's many stakeholders, including DHS; the BioWatch program; the CDC and public health laboratories; state, local, and federal decision makers; and private companies involved in developing BioWatch technology.

The BioWatch program can leverage existing data and guidance to inform the development, testing, and validation of the PCR assays it uses. Relevant recommendations on the development and validation of PCR assays for biothreat agent detection have been developed by the Stakeholder Panel on Agent Detection Assays (SPADA) to support public safety actions and through the Public Health Actionable Assay (PHAA) approach to support public health decision making, efforts that have both been led by DHS's Science and Technology Directorate. Approaches used by other environmental monitoring programs also may be relevant to BioWatch. For example, the U.S. Postal Service operates an automated

Biohazard Detection System aerosol monitoring system to detect *Bacillus anthracis* in mail processing facilities, and the DOD deploys environmental monitoring at selected military locations (e.g., Portal Shield).

None of these assay approaches, however, was developed specifically to support BioWatch's mission. As a result, BioWatch has used a modification of the assay standards recommended by SPADA to test the performance of BioWatch screening and secondary PCR assays. When the program conducted testing of vendor systems as part of efforts to develop autonomous technology,[5] it similarly modified approaches recommended by SPADA to design its testing strategy.

Despite federal multiagency efforts on assay performance and assay performance equivalency, there currently does not appear to be a clear and well-described set of PCR standards, associated validation protocols, and reference materials that are agreed upon and accepted by DHS, the BioWatch program, and the relevant stakeholder partners including the CDC, public health laboratories in the LRN, and those entities involved in developing BioWatch technology.

DHS REQUESTED THAT THE NATIONAL RESEARCH COUNCIL AND THE INSTITUTE OF MEDICINE EXAMINE OPTIONS FOR A BIOWATCH PCR STANDARD

Given the fundamental nature of the BioWatch program's reliance on PCR performance characteristics, the significant investments that the United States makes in BioWatch as a defense against large-scale releases of biological pathogens, and the need for the system's users to interpret and have confidence in its results, DHS asked the National Research Council and the Institute of Medicine to examine the types of performance data, performance criteria, and data analyses that the community requires. A committee appointed by the National Academies was asked to examine PCR performance criteria and validation measures and to evaluate how existing or proposed PCR assay standards meet the program's requirements in terms of performance, cost, and capacity to provide actionable information to leaders in public health. The committee was asked to provide guidance on standards for validation and verification of PCR reactions such that these standards can provide a reasonable measure of confidence to federal, state, and local public health officials and key stakeholders in the BioWatch program. To accomplish this goal, the committee examined current BioWatch protocols and the process used

[5] A proposed "Generation 3" acquisition of autonomous PCR-based detection systems for BioWatch was cancelled by DHS in 2014. See Appendix A for information on the assay and system testing conducted by BioWatch.

to assess the performance of these protocols, and was asked to provide guidance on whether improvements could be made. DHS also requested that the committee consider, to the extent feasible, the implications for PCR assay performance of combining multiple nucleic acid targets into a single PCR (referred to as multiplex PCR). Box 1-2 provides the committee's full statement of task.

There are important limitations in the focus of the report that should be recognized by program stakeholders. A comprehensive PCR assay is a process with multiple steps, only one of which is the PCR amplification. Most of the performance characterization and validation strategies

BOX 1-2
Statement of Task

An ad hoc committee will be convened to conduct a study and prepare a report that will evaluate and provide guidance on appropriate standards for the validation and verification of polymerase chain reaction (PCR) tests and assays in order to ensure that adequate performance data are available to public health and other key decision makers with a sufficient confidence level to facilitate the public health response to a BioWatch Actionable Response (BAR).

Specifically the ad hoc committee will:

1. Determine PCR assay test and evaluation criteria that will provide a reasonable measure of confidence to federal, state, and local public health officials and key stakeholders.
2. Identify and evaluate the Stakeholder Panel on Agent Detection Assays (SPADA), the Public Health Actionable Assays (PHAA), and any other existing and proposed standards applicable for use in defining the performance (validation and verification) of PCR assays for the BioWatch and other programs to ensure confidence as identified in Subtask 1, above. Standards are to be evaluated in terms of performance, cost, and public health applicability.
3. Examine current PCR protocols used by the BioWatch program and other relevant biosurveillance programs and determine if the processes used to assess the performance of these protocols and assays are adequate to meet the standards identified in Subtask 2, above.
4. Determine whether improvements could be made by adopting changes based on the evaluation in Subtasks 2 and 3, above.
5. Determine if any existing standards approach is conducive, taking into consideration cost, schedule, and data requirements, to measuring performance of a PCR assay in multiplexed format.
6. In the event that no approach currently in existence is judged to be appropriate in Subtask 5, above, provide recommendations for aspects that a standard must include to measure performance of multiplex PCR technology.

examined by the committee involve dimensions of the PCR (e.g., the limits of detection of the reaction or the performance of the reaction against inclusivity and exclusivity panels). A detection platform is always associated with the reaction in order to detect the results, and therefore assay performance testing specifies which detection platform is used. However, the report does not compare or recommend one type of detection platform versus another. Additional information on the focus and limitations of the report is detailed in Box 1-3.

BOX 1-3
Limitations of This Report

For a BioWatch sample to make it from an airborne particle to an identified and interpreted signal on a computer leading to an action, it must go through numerous steps, only one of which is the PCR reaction. The input to the PCR reaction is critical to the success and ultimately to the interpretation of the PCR result, as is the way that the PCR amplification process is detected. The thermal cycling conditions and reagents used are also important factors in obtaining a successful PCR product. All of these steps and issues can cause false positive and false negative results in a PCR-based test, even if the PCR reaction is optimized.

The report focuses on performance characterization and validation of the PCR. It comments on aspects of the PCR assay that precede the reaction, including the complications posed by environmental background and the effect of sample preparation. However, the report does not make recommendations on

- The utility or effectiveness of the method of sample collection, such as air filters or air collectors;
- Particular sample processing methodology, including specific methods to extract and purify the nucleic acids from the sample, other than to suggest rigorous validation of all processes;
- How to address the effects of materials that can affect a PCR, such as road or brake dust, pollen, dirt, heavy metals, or other substances, or how to address the effects of background biological organisms and nucleic acids that may or may not be amplified in the PCR process, other than to recommend a rigorous examination of environmental background;
- Particular thermal cycling conditions or instrumentation;
- Specific sources of PCR reagents, or specific methods of storing or dispensing reagents;
- Method of detection of the PCR products, or the size or other particularities of the PCR products; and
- The algorithm for interpreting the results from the PCR.

ORGANIZATION OF THE REPORT

Since its inception, BioWatch has been the subject of a number of reports and congressional hearings that have explored its costs, effectiveness, and roles (e.g., GAO 2009, 2010, 2011, 2012a; IOM and NRC 2011). This report does not address broad programmatic questions on the function of BioWatch and how it fits into the country's public health and security enterprises. Rather, the report discusses performance characteristics of PCR and explores the implications for the BioWatch program and its stakeholders of different types of assay standards approaches. The report seeks to provide guidance to the program to inform decisions made on standards for the PCR assays it uses.

Chapter 1 introduces BioWatch, its mission, and the motivation behind the current study. Chapter 2 discusses the development and use of assay performance standards and provides background information on federal environmental biosurveillance efforts. Chapter 3 explores several of the core components of real-time PCR performance standards and the committee's analysis of tradeoffs arising from different types of standards approaches. Chapter 4 returns to the operational context of the program and the experience of stakeholders in making use of assay results, as well as the important role of a quality assurance program as part of validation. Chapter 5 considers potential impacts on BioWatch of the changing technology landscape, including additional assay performance considerations with multiplexed PCR assays and the implications of continued advances in next-generation sequencing technology. Finally, Chapter 6 summarizes the committee's conclusions and recommendations.

2 Performance Standards Provide Confidence in Results

A laboratory in a BioWatch jurisdiction obtains a positive result from the screening and secondary assays it runs to detect the presence of a biothreat agent. The collector from which the sample came is located in a busy station in the city's mass transit system. A decision needs to be made regarding an actionable response. If officials evacuate the station, significant disruption will result. If they shut down the station for what turns out to be a non-event, they are likely to waste resources, face public anger, affect the local economy, undermine confidence in the city's leadership, and face political consequences. On the other hand, if officials fail to take action for what turns out to be a release of a biological pathogen, the consequences could be far more severe. Therefore, the jurisdiction must know how to interpret the assay results it obtained and what risks are involved with that interpretation.

ESTABLISHING APPROPRIATE PERFORMANCE STANDARDS HELPS RESULTS BE ACTIONABLE

The BioWatch system is substantially more complex than described in the simple scenario above. Nevertheless, experts in the laboratory, along with public health and law enforcement decision makers, need a solid foundation from which to understand assay results and use them in the subsequent decision process. This foundation relies on defining the purpose for which the assay is used, establishing criteria the assay must fulfill to meet its intended purpose, and validating that the assay indeed performs within requirements (or at least to define the assay's limitations). These three pieces—defining purpose, establishing performance standards, and undertaking validation—provide crucial information on which to establish the significance of a result. This chapter and Chapter 3 focus on fundamental characteristics that form the basis for polymerase chain reaction (PCR) assay performance standards. Chapter 4 returns to the validation framework and explores how performance standards fit into the overall picture.

ASSUMPTIONS AND DEFINITIONS FOR "PCR ASSAY" AND RELEVANT STANDARDS

Many terms related to PCR performance characterization and validation have ambiguous definitions in common use. A description of the terms as they are used in the report is provided below.

PCR Assay

There are several ways in which the "PCR assay" step in the BioWatch system could be defined. For example, the assay could be defined to include the set of steps from sample extraction to readout (steps beyond simply thermal PCR amplification cycles and the associated chemistry). The Food and Drug Administration (FDA), for example, generally considers sample extraction in concert with amplification and detection in its device submission guidance. The Environmental Protection Agency (EPA) similarly includes sample collection and extraction in its guidance on PCR method validation because of the impact that these procedures have on the PCR amplification. The BioWatch program, on the other hand, appears to define the PCR assay as starting from an already-extracted sample of nucleic acid (it does not include the collection device or extraction processes), but does include the reagents and analytical platform used for amplification, detection, and data output. The BioWatch program makes protocols available to its jurisdictional laboratories in order to provide standardized methods for the daily operation of sample processing and DNA extraction from BioWatch filters. The program has indicated that it is considering opportunities to improve these protocols, but that such efforts are currently decoupled from the PCR assay test and validation needs that are the subject of the present study. For the purpose of the report, the committee likewise considers the PCR assay to begin with an extracted sample of nucleic acid. It includes the amplification and detection steps along with the required assay reagents and materials, but does not include the downstream decision making undertaken by BioWatch jurisdictions and other local, state, and federal partners following the generation of assay results. In Chapter 4, the committee returns to the impact of the system steps before and after the PCR assay.

Types of PCR Standards

There are multiple types of standards that could be applied to the development and use of PCR assays in BioWatch. These include

- *PCR signature design and maintenance standards:* PCR target sequences, or signatures, must be designed according to the best genomic data available at the time of assay creation. Furthermore, PCR signatures and the assays embodying them are an ongoing process rather than an end point; they must be scrutinized on a regular basis to see if new genomic information will predict false positive and false negative results. Standards guide the approaches used in signature design via *in silico* methods and through selected laboratory testing, and practices for signature maintenance.
- *PCR assay performance standards for validation:*[1] These standards define broad guidelines for a PCR validation strategy, which of necessity must be tailored to some implementation details (e.g., what a "positive detection" means for particular detection chemistry). Independent or multilaboratory validation that the PCR assay meets performance standards is frequently undertaken as a complement to performance characterization work undertaken by the developer.
- *PCR reference collection standards:* PCR assays are tested against inclusivity and exclusivity strain panels (targets, near neighbors, and more distant nontarget organisms) and diverse backgrounds (soils, aerosols, pollens, dust, and other potentially problematic environmental backgrounds). A source of quality-controlled biological materials is needed. Reference standards collections may be provided by an ad hoc consortium or by individual researchers or maintained by a sponsor; access to the materials may be open to government, academia, and industry, or controlled by permission of the owner. In general, a collection of standard reference materials for an application such as BioWatch would include both extracted nucleic acids and live microbial cultures for use by qualified applicants.

---

[1] In this report, the term "validation" refers to the process of affirming that the test or methodology (in this case a BioWatch PCR assay) meets established performance requirements. Validation is also an overarching process or framework, rather than a single step, a concept discussed in further detail in Chapter 4. The term "verification," on the other hand, is a process that affirms that a given user laboratory can obtain the anticipated results and assay performance from a validated PCR assay. These definitions are similar to the FDA's definitions for the terms (FDA 2014c).

- *PCR assay analysis and reporting standards:* How assay results are generated and delivered is important to the BioWatch mission. These types of standards inform the use of a decision algorithm that considers the information provided by relevant controls and the measured result(s) of the PCR assay(s) in the test.
- *End-to-end PCR system performance standards:* Desired sample-in to result-out standards establish performance benchmarks for the overall system, such as the ability to detect a certain number of agents on a filter quadrant with defined overall success rates.
- *PCR proficiency standards:* Those using the BioWatch assays must be well trained in their use. BioWatch user laboratories verify that they can obtain the expected results from the assays they run. To achieve this, the laboratory will likely need to meet specific proficiency standards and undergo regular proficiency testing.

These different forms of PCR standards are interrelated. The committee's task with regard to the BioWatch program focused primarily on the real-time PCR assays used by the program for biological agent detection. As a result, the committee focused particularly on PCR assay performance standards. However, the importance of regular signature maintenance, availability of standard reference collections, and the role of proficiency testing in the overall validation framework are mentioned in the report where appropriate.

Although not discussed in detail, the importance of performance characterization and validation applies equally to multiple processes in BioWatch operations that form part of understanding full system performance. When integrated into a system, several of the preliminary steps will directly impact the PCR assay step. For example, the composition of the collection matrix and airflow characteristics may influence how agents are deposited and recovered,[2] while the procedures used for extracting DNA from the filter directly affect the amount and quality of the material available for input to the PCR. Within an instrument, different methods for detecting and interpreting the signal from the PCR amplification can affect the assay results and performance for downstream decision-making algorithms. As Figure 2-1 emphasizes, these activities are interconnected within the BioWatch program and it is necessarily somewhat artificial to isolate the PCR assay step. Full system operational performance testing and the implementation of quality assurance standards for the program, discussed further in Chapter 4, do take into account additional steps from filter analysis.

[2] Currently, the BioWatch system makes use of aerosol deposition onto dry filters, but other types of collection approaches could be used, such as liquid-based collection.

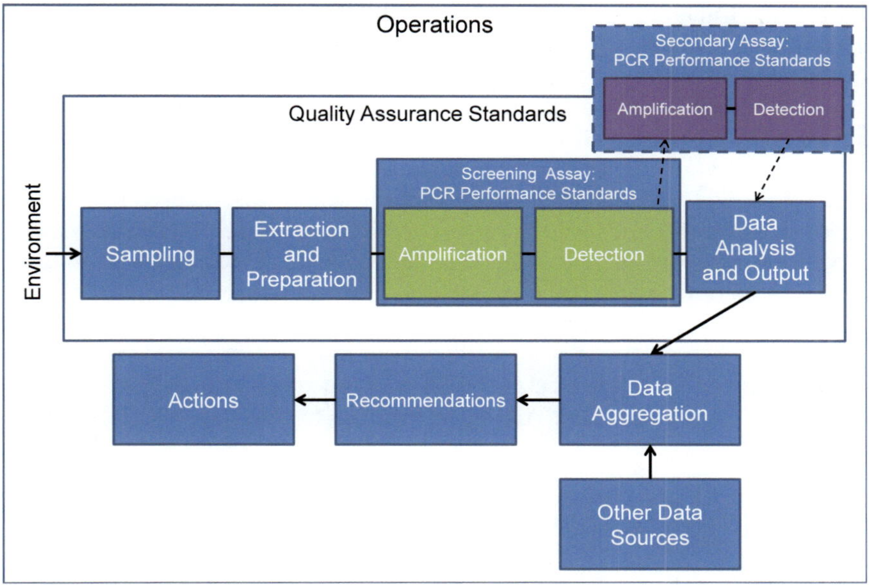

FIGURE 2-1 BioWatch operations include multiple steps. PCR assay performance standards are associated with both the screening and secondary assays, and quality assurance standards are associated with the framework from sampling through output.

DEVELOPING AND USING PERFORMANCE STANDARDS

Guidance on the development and use of performance standards by the federal government is provided by the National Technology Transfer and Advancement Act of 1995 (NTTAA; P.L. 104-113) and Office of Management and Budget (OMB) Circular A-119 on Federal Participation in the Development and Use of Voluntary Consensus Standards and in Conformity Assessment Activities (OMB 1998[3]). These guidelines encourage agencies to make use of voluntary consensus standards and to participate in voluntary consensus standards efforts where they are relevant to the agency's mission. The guidance reserves the right of agencies to use other types of standards, such as government-unique standards, where use of standards developed by a voluntary consensus body would be "inconsistent with applicable law or otherwise impractical" (NTTAA, § 12(d), 15 USC § 272).

[3] The federal government currently is considering revisions to OMB A-119. See http://www.whitehouse.gov/sites/default/files/omb/inforeg/revisions-to-a-119-for-public-comments.pdf.

Under the Homeland Security Act of 2002 (P.L. 107-296), the Department of Homeland Security (DHS) has been directed to conduct its standards activities in accordance with the NTTAA and OMB A-119.

In its analysis, the committee explored several assay performance approaches, some created by voluntary consensus bodies and published in the open literature and others considered to be government-unique and containing information restricted from public disclosure. The committee does not attempt to distinguish which are or are not performance standards under a formal definition of the term, but rather comments on the types of information they include and the potential tradeoffs and implications they present for BioWatch. For BioWatch to provide sufficient confidence to its users, however, some mechanism for engaging the participation of its relevant stakeholders is required, many of whom are outside of federal agencies. The report returns to this issue in Chapter 4.

PCR ASSAY PERFORMANCE AND VALIDATION APPROACHES THAT MAY BE APPLICABLE TO BIOWATCH

By 2005, several nonclinical surveillance programs were operational to detect selected biological pathogens. These included the BioWatch program, the Biohazard Detection System (BDS) used by the Postal Service to screen mail at major processing centers, and Department of Defense (DOD) programs such as the Guardian Installation Protection program to monitor specified U.S. and international military facilities. Two performance standards, in particular, also have since been developed that may be applicable for use or adaptation by the BioWatch program to validate the performance of the real-time PCR assays it employs. These are the performance requirements developed under the Public Health Actionable Assays (PHAA) program and the Standard Method Performance Requirements and Validation Guidelines established by the Stakeholder Panel on Agent Detection Assays (SPADA).

Public Health Actionable Assays

Under the leadership of the Science and Technology Directorate, DHS established a program to develop performance requirements for assays to detect specific microbial agents as Public Health Actionable Assays. The program was established in 2006 to support the goals of a national biomonitoring architecture called for in an interagency memorandum of understanding on monitoring for threat agents that had been signed in 2005. The original PHAA program subsequently split, with one piece continuing as PHAA and another as a Public Safety Actionable Assays (PSAA) program.

Under PHAA, DHS and the Centers for Disease Control and Prevention (CDC) consider such "public health actionable assays" to support high-confidence agent detection and characterization performed in a laboratory setting, and to support public health actions such as deployment of antibiotics. The PHAA program was built on and expanded the strategies used for testing assays in the Laboratory Response Network (LRN) (Morse 2014). The PHAA process involved participation of subject-matter experts from federal agencies and the academic community to develop recommended inclusivity and exclusivity panels and other aspects of the assay evaluation approach. In contrast to SPADA, PHAA was developed to be a government-unique standard with a focus on government-operated programs; details of the discussions and performance requirements are not disseminated in the open literature (DHS 2014; Pillai 2014).

The PHAA documents describe reference materials and testing approaches to be used for both clinical samples and environmental samples. Assays to detect environmental samples under PHAA, the subset that would be relevant to BioWatch, require validation testing with a different subset of exclusivity strain panels than clinical samples (e.g., they do not include the PHAA panel representing clinical background microorganisms). The CDC signed on to the PHAA standard in 2011 for use with its agent detection assays through the LRN (Merlin 2014).

Stakeholder Panel on Agent Detection Assays

SPADA operates under the auspices of AOAC International, an association of analytical communities, as a voluntary consensus standards body and is supported by funding from DHS. It engages the participation of over 100 representatives from federal, state, and local governments, the first-response and public health communities, academia, and industry. SPADA has established standards for real-time PCR methodologies to detect the pathogens *Francisella tularensis*, *Yersinia pestis*, *Bacillus anthracis*, *Burkholderia pseudomallei*, and *Burkholderia mallei* from aerosol (AOAC 2011a,b,c; AOAC in preparation-a,-b) and is currently working to develop standards for Variola virus (AOAC 2014). The standards are based on panels of recommended inclusivity and exclusivity strains, environmental substances, and validation test protocols. In addition to standards for PCR-based assays to detect aerosolized agents, SPADA has established standards for immuno-based handheld assays to screen suspicious powders for use by first responders (AOAC 2011d,e). The SPADA process for establishing these performance standards relies on 10- to 20-person working groups of subject-matter experts, who develop recommendations for discussion by the full membership of SPADA and voting by approximately 20-25 SPADA voting members. Once a standard has been approved, it is publicly published in the *Journal of AOAC International* (Coates et al. 2011; Davenport 2014).

The performance requirements established by SPADA are considered by DHS and CDC to be Public Safety Actionable Assays, which are suitable for first responders in the field to use to support incident management, including building evacuation. Samples testing positive with a PSAA standard would subsequently be transported to the CDC LRN for further analysis.

Department of Defense Assay Transition Packages

The DOD also makes use of nucleic acid–based detection technology to support biosurveillance on military facilities, as well as other types of detection methods (e.g., protein based) and sensors. The approach taken by DOD is to standardize essential information that is captured and presented during assay development to sufficiently document performance. This process is defined in a Defense Technology Objective (DTO) CB.56 molecular assay transition package, a format that was developed by multiple DOD stakeholders and began to be used across the agency in 2005 (Goodwin 2014).[4] According to DOD, "[a]n enabling technology DTO focuses on the development of knowledge to address a specific issue, and is a necessary intermediate step to achieve an operational capability" (DOD 2005). Thus, documentation such as the molecular assay transition packages would be part of the required process for transition into advanced development. The concept of the DTO CB.56 data transition package and the categories of information it includes have subsequently been recognized by the FDA as the basis for initial submission for pre–emergency use authorization (EUA) and as the framework for assay development and performance information by international DOD partners in Canada, the United Kingdom, and Australia (Emanuel 2014).[5]

For real-time PCR assays, the information contained in the DTO molecular assay transition packages is similar to guidance that has been described by the Minimum Information for Publication of Quantitative Real-Time PCR Experiments (Bustin et al. 2009), which were developed to harmonize descriptions of PCR assays reported in scientific journals. The DOD package generally includes such information as

[4] Stakeholders involved in the discussion included the U.S. Army Medical Research Institute of Infectious Diseases, U.S. Army Dugway Proving Ground, U.S. Army Edgewood Chemical Biological Center, U.S. Navy, Joint Biological Agent Identification and Diagnostic System, regulators, and developers (Goodwin 2014).

[5] For example, an EUA was recently issued by the FDA for *in vitro* diagnostic use of a real-time PCR Ebola virus detection assay developed within DOD and documented by a DTO CB.56 information package (FDA 2014b).

- Description of target with database accession numbers, gene location, and sequence;
- Database searches of target gene and sequences producing significant alignment (with accession numbers, names, and percent similarity);
- Software used to design primers and probe, primer and probe sequences and locations, characteristics such as length, melting temperature, and percent GC content, and information on database sequences producing significant alignment with primers and probe sequences;
- Assay reaction conditions including information on the specific reagents and instruments used and the thermal cycling parameters;
- Information on optimizing assay conditions (such as testing different primer or probe concentrations and the resulting threshold cycle [C_t] values and end point fluorescence);
- Calculations of assay linearity (and associated data such as equation, slope, and R^2 value);
- Determination of assay limit of detection; and
- Information on assay inclusivity and exclusivity revealed by testing against various strains.

The committee's understanding is that over time, DOD assay developers have created a pool of potential pathogen detection assays, reagents, and the accompanying packages that characterize their performance, so that missions may select suitable assays as the need dictates. When an assay is selected and tested for a particular mission need, the committee assumes that the performance of the assay and system are validated in the context of the operational setting. The committee did not receive details on how DOD conducts performance and validation testing beyond dimensions described in the DTO CB.56 package when the assay is developed.

Within DOD, the Critical Reagents Program (CRP) serves as a consolidation point for quality control programs, standard operating procedures, conformance test plans, and assay reference materials. The CRP credits sustained funding through DOD 5-year budget cycles for successful long-term operation and maintenance of this resource (Goodwin 2014).

Other Federal Programs

The BDS is an autonomous air sampling and analysis system designed to detect the presence of *Bacillus anthracis* in postal facilities through PCR-based detection. If a positive sample is recorded, material is sent to the CDC's LRN for confirmatory testing (O'Neill 2014). The EPA also provides guidance on validation for microbiological methods (EPA 2009). The guidance provided by the EPA is similar in nature to the performance guidance provided by SPADA, PHAA, DOD, and FDA. The EPA requires a written description of the method by the developing laboratory, summarizing the method; describing the equipment, reagents, and standards used; providing information on quality control measures, data, and statistical analysis; giving method performance results; and indicating other details. Within-laboratory performance parameters such as assay sensitivity, specificity, and precision (also referred to by EPA guidance as "primary validation") are reported. The EPA takes a tiered approach to the nature and extent of multilaboratory validation required, based on the use of the assay (with a focus on conducting more extensive validation for methods that will, for example, be used nationally). The EPA guidance notes that different detection chemistries will influence PCR performance and that assays may perform differently on different types of instruments. As a result, validation is necessary in the context of the instrument and chemistry to be used, with appropriate documentation and additional validation likely to be necessary if these aspects change.

PRIOR FEDERAL EFFORTS ON PERFORMANCE STANDARDS AND ASSAY EQUIVALENCY

Given the fundamental role of performance standards and validation, the widespread use of PCR-based assays, and the increasing emphasis on biosurveillance since 2001, it is not surprising that federal agencies have a history of discussing these topics. Table 2-1 highlights several of the relevant events that have occurred.

One set of efforts has been directed at the issue of establishing federal assay performance equivalency. Multiple agencies and programs were using similar types of assays, and there was interest in having a common basis for comparing performance. In 2005, a Memorandum of Understanding (MOU) on Coordinated Monitoring of Biological Threat Agents was signed by DHS, the Department of Health and Human Services (HHS), DOD, the U.S. Postal Service (USPS), and the Department of Justice (DOJ) (Appendix A in DHS 2011). The agreement discussed development of a basis to demonstrate performance equivalency among the PCR assays used by federally owned biomonitoring activities to confirm the presence of biothreat agents. The text of the MOU, which expired on January 1, 2010, reads in part,

> By August 2005, the Participants will use biological threat agent detection assays, protocols, and response algorithms for confirmation of a biological threat agent (presence of biothreat agent nucleic acid) that are

TABLE 2-1 Time Line of Selected Programs and Milestones

Year	Program	Milestones
1997	DOD Critical Reagents Program established	Provide a source of quality-controlled reference materials, reagent, and assays to support the missions of the DOD and its partners; CRP subsequently brings repositories under ISO 17025
1999	CDC Laboratory Response Network established	Provide a network of state and local public health, veterinary, military, and international laboratories to respond to biological, chemical, and radiological threats and other public health emergencies
2001	BASIS (Biological Aerosol Sentry and Information System) developed for biosecurity at 2002 Winter Olympics	Pressed into daily use in October 2001 following mailings of letters containing *Bacillus anthracis*; later expanded to become BioWatch
2003	DHS BioWatch Program established	Provide early warning detection of significant releases of selected pathogens in major metropolitan areas and for special events
2005	DOD Test and Evaluation Capabilities and Methodologies Integrated Process Team (TECMIPT) established	Established under Chemical and Biological Defense Program (CBDP) Test & Evaluation (T&E) Executive to provide recommendations on T&E strategies and standards
2005	MOU on Coordinated Monitoring of Biological Threat Agents by DHS, HHS, DOD, USPS, and DOJ.	Calls for effort on test performance equivalency for confirmation of a biothreat agent, to be coordinated by DHS
2005	BioNet PCR Equivalency Study	Goal of demonstrating assay performance equivalency between deployed DOD assays and CDC LRN assays for environmental monitoring programs
2005	DTO CB.56 Molecular Assay Transition package begins to be used across DOD	Provide a harmonized set of descriptive information and performance details about assays developed for use by DOD; subsequently agreed-upon as assay development information by international partners under Chemical, Biological, and Radiological (CBR) MOU
2006	DHS establishes Public Health Actionable Assay program	Established in support of the goals in the biomonitoring MOU

continued

TABLE 2-1 Continued

Year	Program	Milestones
2007	Stakeholder Panel on Agent Detection Assays established through AOAC International	Voluntary consensus standards body with participants from federal, state, and local governments, the first-response and public health communities, and industry to develop performance and validation requirements for evaluation of environmental detection technologies
2007-2009	Introduction and subsequent rollback of multiplex assays by BioWatch	
2008	Original PHAA program splits into Public Safety Actionable Assay (PSAA) and PHAA programs	PSAA program focuses on first responders and private sector; SPADA considered PSAA; PHAA focuses on federal biodetection programs
2010	DTO CB.56 assay packages accepted by FDA as initial pre-EUA submissions	Emanuel (2014)
2011	NSTC Committee on Homeland and National Security Subcommittee on CBRNE Standards established (SOS-CBRNE)	Provide forum to communicate and coordinate on relevant standards and technologies. Replaces previous NSTC subcommittees on standards and on decontamination standards and technology
2011	CDC LRN signs on to PHAA Standard	Merlin (2014)
2011	BioWatch program transitions to using CRP-provided materials for its agent screening assays	Walter (2014b)
2013	SPADA working group on performance requirements for Variola PCR assay established	SMPR released in late 2014
2014	DHS cancels the BioWatch "Gen-3" acquisition program	
Ongoing	Assembly of PHAA strain panels being undertaken	
Ongoing	CDC submitting LRN PCR assays for FDA approval (for clinical use)	

> mutually acceptable to the Participants. . . . Using predetermined performance criteria, the Participants will evaluate rapid biological threat detection assays (e.g., PCR or other technologies), protocols, and response algorithms on select agents, similar substances, and other interferents to support a determination of equivalency. (Appendix A, lines 154-166, in DHS 2011)

The committee notes that the MOU refers to method equivalency for *confirmation* (emphasis added) of agent detection.

A "BioNet" study was undertaken between PCR assays deployed by DOD and those used by the CDC LRN. The multiyear, multi-million-dollar effort involved developing an agreed-upon test plan and comparing the performance of real-time PCR assays then in use by the CDC LRN, DOD Joint Program Executive Office for Chemical and Biological Defense, and DOD Naval Medical Research Center. Strain DNA was provided to Lawrence Livermore National Laboratory, which prepared 96 well plates of blinded samples that were sent to Battelle Memorial Institute for independent testing. Assays developed by CDC and DOD to detect three different pathogens of potential relevance to BioWatch were evaluated. Each assay's limit of detection was characterized using a reference strain and assay specificity was characterized using panels of inclusivity, exclusivity, and background organisms, as well as background matrix provided by BioWatch and DOD Guardian air filters (Joan Gebhardt, Naval Medical Research Center [NMRC], personal communication, January 6, 2015).[6] The strain panels used for testing were based on the strains obtainable; more limited strain typing data were available at the time, and thus it is not clear how well the strains available covered the genetic diversity of the tested pathogens (Segaran Pillai, DHS, personal communication, January 21, 2015).

To the best of the committee's understanding, the report resulting from BioNet indicated that the CDC and DOD assays that were tested performed equivalently based on the collected data (Joan Gebhardt, NMRC, personal communication). On the other hand, the BioNet project appears not to have succeeded in a long-term, agreed-upon basis for establishing assay equivalency (Emanuel 2014). An ongoing effort under the leadership of the DHS Science and Technology Directorate (S&T) is now focused on establishing Federal Standards for Evaluating Assay Performance and Equivalency (FSAPE). The objective of this effort is to "expedite consequence management efforts responding to intention-

[6] For each of the three tested pathogens, panels of approximately 30 inclusivity strains were assembled (tested at 10 pg of DNA) along with panels of varying numbers of near-neighbor exclusivity strains and a panel of background organisms (tested at 100 pg of DNA); panel composition was based on the strains that the effort was able to obtain.

ally released pathogens that have been detected by federally owned and operated biomonitoring programs. . . . This improved understanding will allow for interoperability, which is essential for coordinated consequence management efforts across agencies and to support public health actions and decisions (DHS 2011, p. 9). According to Segaran Pillai, Chief Medical and Science Advisor in DHS S&T, the development and implementation of FSAPE is being undertaken in collaboration with federal agencies such as DHS, DOD (Office of the Secretary of Defense), HHS (Office of the Assistant Secretary of Preparedness and Response, CDC, FDA), DOJ (Federal Bureau of Investigation), USPS, U.S. Secret Service, and state and local public health leaders and associations such as the Association for State and Territorial Health Officials, Association of Public Health Laboratories, and the National Association of County and City Health Officials (Pillai 2014).

The FSAPE document describes performance criteria and validation protocols that are to be used to document assays. These criteria and protocols are essentially identical in approach to the PHAA standard for validating assays from environmental (nonclinical) samples and so PHAA and FSAPE are discussed together where relevant in the remainder of the report. FSAPE appears to be an ongoing effort, and it is not clear to the committee whether all of the relevant federal agencies have formally signed on to this approach.

ASSAY PERFORMANCE CHARACTERIZATION CONDUCTED BY LOS ALAMOS NATIONAL LABORATORY

On behalf of the BioWatch program, Los Alamos National Laboratory (LANL) assembled and quality-controlled strain panel materials, developed a test plan to characterize the analytical performance of several of the assays being used by the program, and compared assay performance on vendor and reference systems to inform the BioWatch Gen-3 acquisition program on autonomous detection capabilities (LANL 2010, 2011a,b, 2012a,b). The LANL work partially followed SPADA guidelines, with some alterations. LANL used purified DNA for its analytical performance testing. Subsequent BioWatch testing that included extraction from a filter matrix was conducted at Dugway Proving Ground as part of full system testing. The discussion in the report is informed by approaches such as SPADA and PHAA/FSAPE, as well as by the procedures that LANL employed. DHS also has undertaken system testing beyond assay analytical performance. Table 2-2 summarizes recent activities.

Prior federal and stakeholder efforts on PCR assay performance standards provide a basis for discussing the tradeoffs of different approaches that may be applicable to BioWatch. A discussion of critical components of

TABLE 2-2 Recent BioWatch Performance Testing and Validation Activities[a]

Year	Event or Activity
2011	BioWatch Quality Assurance program established
2012	BioWatch assay performance characterization and evaluation undertaken by Los Alamos National Laboratory
2013	BioWatch reference system characterization test undertaken at Dugway Proving Ground
2014	Field test of current BioWatch (Gen-2) system to be conducted at Naval Surface Warfare Center, with goal of testing system detection in an operationally relevant environment

[a]A restricted version of this table may be found in Appendix A.

performance standards is presented in Chapter 3. The committee returns in Chapter 4 to the broader validation framework, the important role of monitoring and feedback, and the operational context of assay use. Although BioWatch is owned by a federal agency, it is managed and operated locally; it is the experience and confidence in results of the users in the public health laboratories and the jurisdictional decision makers that are critical to the success of BioWatch.

3 Key Components of a Performance Standard for BioWatch

This chapter explores fundamental components of performance standards for real-time polymerase chain reaction (PCR) assays and discusses basic principles that a pathogen detection standard would need to take into account to serve the BioWatch program's needs. The chapter compares approaches taken by existing guidelines, as well as performance characterization work undertaken at Los Alamos National Laboratory (LANL), and provides committee suggestions.

The committee determined that a PCR standard to characterize assay performance and conduct validation, applicable to BioWatch, would need to meet the following basic goals:

- Establish that the assay has a sufficient limit of detection and sensitivity to detect the release of a tested biothreat agent at a program-relevant concentration above the baseline of environmental background. The BioWatch system was not deployed with the goal of detecting a single airborne particle, but rather to monitor for a substantial release event. On the other hand, low concentrations of agents must be detectable given the variability introduced by wind dispersion and collector siting, as well as losses and inefficiencies in the subsequent analytical steps leading to the PCR reaction, should there be an environmental pathogen plume.
- Establish that the assay has sufficient specificity to not cross-react with nontargeted strains and background organisms or materials, within acceptable program false positive and false negative rates.

• Be sufficiently robust for routine operational use by BioWatch jurisdictional laboratories and be otherwise acceptable to program users.

The simple flowchart in Figure 3-1 indicates some of the main steps in the process of establishing whether assay characteristics meet these goals.

TRADEOFFS IN PERFORMANCE AND ASSAY PURPOSE

Under the current BioWatch program, an initial screening assay (one signature per assay; assay and materials provided through Department of Defense [DOD] Critical Reagents Program [CRP]) is followed by a secondary assay (three or more signatures in the assay set; assay and materials provided through Centers for Disease Control and Prevention [CDC] Laboratory Response Network [LRN]) if the initial assay produces a positive result.[1] There is an inherent tradeoff between the properties of sensitivity and specificity—as the assay's true positive rate increases so does the false positive rate. Therefore the specificity of the assay decreases. A screening assay should prioritize sensitivity. If a nucleic acid from a tested pathogen is present then the assay should detect it and minimize the rate of obtaining a false negative result. A false positive result is less detrimental at this stage because all positive screening results will proceed to a subsequent secondary assay. A secondary assay for verification or confirmation, on the other hand, should have high specificity. The assay needs to be able to precisely distinguish between the tested pathogen and closely related but nonpathogenic organisms that the assay should not detect, minimizing the rate of a false positive result being passed forward for action. While this type of approach has been the general compromise between the two assays, the secondary assay should and likely can be comparable in sensitivity with that of the screening assay. The analytical system must be able to achieve a target limit of detection (LOD) to minimize a situation in which the primary assay yields a positive result but the secondary assay produces a false negative that causes the primary result to be discounted. The combination of the initial screening and secondary assays function together to provide an overall high sensitivity and high specificity. Several participants in BioWatch jurisdictions from whom the committee heard during its data-gathering indicated that the program's evolution to using a

[1] The signature used in the initial screening assay is not necessarily the same as the set of signatures used in the secondary assay, although there may be places where these signatures overlap for some tested agents. Both screening and secondary assays could be provided through the DOD CRP, in which case the committee understands that the secondary assay would repeat the screening signature. The program currently uses the LRN secondary assay, which targets a different set of signatures.

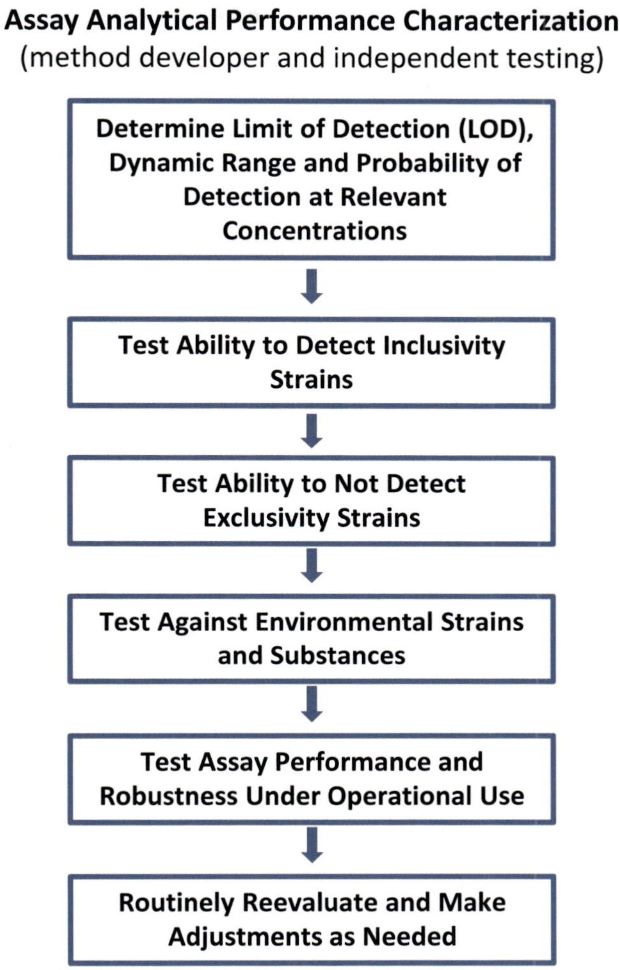

FIGURE 3-1 Flowchart representing steps in assay analytical performance characterization.

two-step assay process has improved their experiences with regard to preliminary positive results and downstream decision making.

Although the same core performance standard and validation framework is appropriate for both types of PCR assays, the BioWatch program could consider the use of serial testing or "tiers" of validation testing that emphasize different parameters based on the assay's actual intended use. For example, stringent characterization of the minimum concentration of inclusivity strains that can be reliably detected with a sufficiently high probability (assay sensitivity) could be prioritized during development and validation of a screening assay along with a basic set of inclusivity, exclusivity, and environmental strains. Because of the importance of specificity to the performance of a secondary assay, the program might conclude that it is useful, for example, to conduct laboratory validation using additional inclusivity and exclusivity strains than were used for an initial screening assay. An assay meant to be used for a secondary level of verification or confirmation could thus be validated in the laboratory against a second tier containing additional inclusivity, exclusivity, and environmental strains in order to demonstrate the requirement that this type of assay have high specificity.

The report focuses the majority of its discussions on the current screening/secondary assay model used by BioWatch for the practical reason that this is the current approach and is likely to remain so for at least the near future. However, the report raises other possible options, particularly for secondary assays, both here and in Chapter 5, for consideration by BioWatch and its network of stakeholders.

ASSAY SENSITIVITY AND LIMIT OF DETECTION

To be suitable for BioWatch's mission, a performance standard must specify the process for determining or confirming that the assay's sensitivity meets program targets. These program targets are (or would be) set by the Department of Homeland Security (DHS). Although very challenging to quantify, the BioWatch program could consider establishing minimum target LODs for its agent detection assays based on factors such as the scale of aerosol release that the program is designed to detect, modeling of plume dispersion, collector siting, collector intake rate, and other parameters that influence the deposition of agent particles on a filter, infectious dose of the pathogens of interest, and system testing characterizing expected losses from the point of release to the filter and through the sample preparation steps that provide input DNA to the PCR amplification. Should such program targets exist, the validated LOD for a given assay would thus be required to meet or exceed (be more sensitive than) the desired LOD in order for the assay to be judged fit for the purpose for which BioWatch is designed. A "notional example" of system losses at various steps is provided in Appendix A. The committee does not have information on the specific performance of the BioWatch system or on operational program LOD targets (e.g., ability to detect x number of copies of DNA for agent A per filter quarter). As a result, the discussion below on the process for validating assay LOD is presented in more general terms.

Assay sensitivity generally is characterized by determining the probability of detection at one or more nucleic acid concentrations. The approach taken by the Stakeholder Panel on Agent Detection Assays (SPADA) in its Standard Method Performance Requirements (SMPRs) and validation guidelines is interesting because it uses a predetermined acceptable minimum detection level (AMDL) of a reference strain. The AMDL value is specified as part of the SMPR, although the SMPR does not provide detailed information on how the specific value was determined. For the detection of *Bacillus anthracis* in aerosol collection filters and/or liquids, for example, the AMDL prescribed by the standard is "20,000 standardized *Bacillus anthracis* Ames spores per filter; 2000 standardized spores per mL; 2000 genome equivalents per mL" (AOAC 2011c). As depicted in Figure 3-2, the SPADA probability of detection (POD) at the AMDL is related to other commonly used assay performance terms.

Other approaches to characterizing assay performance (e.g., the Public Health Actionable Assay [PHAA] and Federal Standards for Assay Performance and Equivalency [FSAPE] guidelines, DOD assay transition packages, and test and evaluation work conducted for BioWatch by LANL) do not start with a set AMDL. Rather, they call for the use of serial dilutions of a reference strain to determine the ability of the assay to yield a positive detection over a range of concentrations. These data are used to calculate the LOD of the assay.

Setting an upfront AMDL performance target has the advantage of providing clear guidance to assay and technology developers on the detection limit that is required for the assay to be acceptable. This model may be more challenging to implement in practice—it is difficult to draw definitive links between the human infectious dose of a pathogen, the pathogen concentration in an air plume that would be required to cause illness, the factors that influence pathogen deposition onto an aerosol point collector like the BioWatch devices, and the effects of collection medium and sample processing prior to entry into the PCR reaction in order to set the appropriate AMDL. Other assay performance approaches essentially have taken an alternate approach at the assay evaluation stage by assuming that an assay will be designed to be as sensitive as possible and determining the achievable lower analytical detection limit. Subsequent decision making is required to determine whether the measured performance meets mission needs. Either approach—using a predetermined AMDL if stakeholder agreement on this value can be obtained or calculating minimum nucleic acid quantity detectable by the assay followed by stakeholder determination of whether this makes the assay acceptable for the mission—seems reasonable to the committee.

Although pathogen detection assays for environmental biosurveillance programs collectively are expected to detect multiple inclusivity

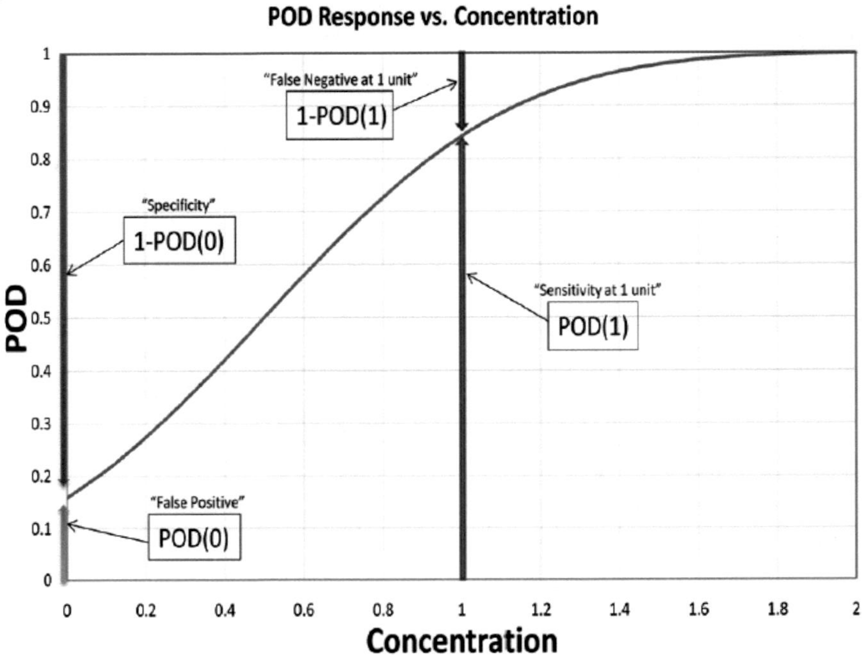

FIGURE 3-2 A hypothetical POD curve, as defined by the SPADA validation guidelines, depends on concentration and can be related to other commonly used assay performance concepts. According to AOAC, "[i]n other models, the terms "false positive," "false negative," "sensitivity," and "specificity" have been defined in a variety of ways, usually not conditional on concentration. For these reasons, their use is denigrated under this model" (AOAC Methods Committee 2011, p. 1374).
SOURCE: AOAC Methods Committee (2011). Reprinted with permission from the *Journal of AOAC International.*

strains of the tested pathogen, it would be cost and time prohibitive to test multiple replicates of all strains to comprehensively determine the LOD for each. Most performance characterization methods use a reference strain to evaluate the assay's performance (whether at a preset AMDL or a calculated LOD). The reference strain method assumes that the LOD will be similar among inclusivity strains that the assay should detect. This assumption is not necessarily accurate. When conducting assay testing using SPADA strain panels, for example, LANL scientists observed that different strains exhibited different probabilities of detection at the anticipated LOD. LANL and the BioWatch program addressed this issue by conducting subsequent steps of testing (for inclusivity and exclusivity panel performance) at a concentration that would allow for some variability.[2] The committee believes that the reference strain approach is a reasonable first step given practical constraints on testing, but variation in true LOD among strains is a potential issue of which programs making operational use of such assays should be aware. The committee suggests that a limited number of reference strains (such as n = 3-5) should be tested in a range of concentrations near the expected LOD in order to get an appreciation of inherent variation.

Multiple replicates must be run to determine the probability of detection at any given concentration with associated statistical information on the confidence limits around the result. Table 3-1 provides examples of confidence limits associated with different numbers of replicates and different numbers of "successes" calculated using the Agresti-Coull, Clopper-Pearson, and Wilson confidence interval methods for binomial data (e.g., detect or nondetect results) using a 90 percent confidence interval (95 percent upper confidence limit and 95 percent lower confidence limit). Although the results are similar for the three methods used, they are not identical and all three are presented here for comparison.[3]

For example, if 10 replicates of DNA are tested and an assay successfully detects DNA in 8 of the 10 samples, there is 95 percent confidence that the true probability of detection of the assay at that concentration is at least ~0.50. On the other hand, should DNA be detected in 40/40 samples, the probability of detection at that concentration is at least ~0.93 (93 percent) with 95 percent confidence. As can be seen in Table 3-1, depending on the interval calculation method employed, testing 47-56 samples with zero failures, 79-85 samples with one failure, and 107-112 samples with two failures will achieve probabilities of detection of approximately 95 percent with 95 percent confidence.

---

[2] For inclusivity strain testing, testing was conducted not at the concentration for which the assay yielded 95 percent probability of detection of the agent reference strain, but at the higher agent concentration for which the assay yielded a 99.5 percent probability of detection of the reference strain. Exclusivity testing was conducted at 10× the 99.5 percent probability of detection concentration.

[3] Current BioWatch assays rely on real-time PCR, which provides an appreciation of the relative amount of nucleic acid present in the sample through differences in cycle threshold values. However, the decision algorithms (employing cutoff C_t values) essentially convert the data to binomial yes/no detection results. The Wilson method is referenced in AOAC validation guidelines (AOAC Methods Committee 2011) while the Clopper-Pearson method is used by Signature Science, LLC in several analyses it has conducted on behalf of DHS as part of the Quality Assurance Program (DHS 2013a,b) because it provides a more conservative estimate of confidence intervals. A prior analysis of bionomial confidence intervals recommended the Wilson method for smaller n and the Agresti-Coull method for larger n (Brown et al. 2001).

TABLE 3-1 Comparison of Binomial Confidence Intervals for Varying Numbers of Tests, Calculated Using Several Confidence Interval Methods

x	n	Agresti-Coull			Clopper-Pearson			Wilson		
"successes"	trials	lower		upper	lower		upper	lower		upper
3	3	0.470	≤ p ≤	1.000	0.368	≤ p ≤	1.000	0.526	≤ p ≤	1.000
3	5	0.271	≤ p ≤	0.859	0.189	≤ p ≤	0.924	0.272	≤ p ≤	0.857
4	5	0.422	≤ p ≤	0.968	0.343	≤ p ≤	0.990	0.435	≤ p ≤	0.954
5	5	0.599	≤ p ≤	1.000	0.549	≤ p ≤	1.000	0.649	≤ p ≤	1.000
9	9	0.731	≤ p ≤	1.000	0.717	≤ p ≤	1.000	0.769	≤ p ≤	1.000
8	10	0.533	≤ p ≤	0.939	0.493	≤ p ≤	0.963	0.541	≤ p ≤	0.931
9	10	0.636	≤ p ≤	0.994	0.606	≤ p ≤	0.995	0.652	≤ p ≤	0.977
10	10	0.751	≤ p ≤	1.000	0.741	≤ p ≤	1.000	0.787	≤ p ≤	1.000
18	20	0.730	≤ p ≤	0.975	0.717	≤ p ≤	0.982	0.738	≤ p ≤	0.966
19	20	0.791	≤ p ≤	1.000	0.784	≤ p ≤	0.997	0.804	≤ p ≤	0.989
20	20	0.859	≤ p ≤	1.000	0.861	≤ p ≤	1.000	0.881	≤ p ≤	1.000
38	40	0.854	≤ p ≤	0.989	0.851	≤ p ≤	0.991	0.860	≤ p ≤	0.983
39	40	0.887	≤ p ≤	1.000	0.887	≤ p ≤	0.999	0.895	≤ p ≤	0.994
40	40	0.924	≤ p ≤	1.000	0.928	≤ p ≤	1.000	0.937	≤ p ≤	1.000
58	60	0.900	≤ p ≤	0.993	0.899	≤ p ≤	0.994	0.904	≤ p ≤	0.989
59	60	0.923	≤ p ≤	1.000	0.923	≤ p ≤	0.999	0.929	≤ p ≤	0.996
60	60	0.948	≤ p ≤	1.000	0.951	≤ p ≤	1.000	0.957	≤ p ≤	1.000
94	96	0.936	≤ p ≤	0.996	0.936	≤ p ≤	0.996	0.939	≤ p ≤	0.993
95	96	0.951	≤ p ≤	1.000	0.952	≤ p ≤	0.999	0.955	≤ p ≤	0.998
96	96	0.967	≤ p ≤	1.000	0.969	≤ p ≤	1.000	0.973	≤ p ≤	1.000
118	120	0.948	≤ p ≤	0.997	0.948	≤ p ≤	0.997	0.951	≤ p ≤	0.994
119	120	0.960	≤ p ≤	1.000	0.961	≤ p ≤	1.000	0.964	≤ p ≤	0.998
120	120	0.973	≤ p ≤	1.000	0.975	≤ p ≤	1.000	0.978	≤ p ≤	1.000
47	47	0.935	≤ p ≤	1.000	0.938	≤ p ≤	1.000	0.946	≤ p ≤	1.000
53	53	0.942	≤ p ≤	1.000	0.945	≤ p ≤	1.000	0.951	≤ p ≤	1.000
56	56	0.945	≤ p ≤	1.000	0.948	≤ p ≤	1.000	0.954	≤ p ≤	1.000
57	57	0.946	≤ p ≤	1.000	0.949	≤ p ≤	1.000	0.955	≤ p ≤	1.000
78	79	0.941	≤ p ≤	1.000	0.941	≤ p ≤	0.999	0.945	≤ p ≤	0.997
84	85	0.945	≤ p ≤	1.000	0.945	≤ p ≤	0.999	0.949	≤ p ≤	0.997
85	86	0.945	≤ p ≤	1.000	0.946	≤ p ≤	0.999	0.950	≤ p ≤	0.997
105	107	0.942	≤ p ≤	0.996	0.942	≤ p ≤	0.997	0.945	≤ p ≤	0.994
110	112	0.945	≤ p ≤	0.997	0.945	≤ p ≤	0.997	0.947	≤ p ≤	0.994

Notes: Confidence intervals calculated using the package 'bionom' version 1.1-1 (Dorai-Raj 2014) in the statistical computing software R, version 3.1.2 (R Project for Statistical Computing, available at http://www.r-project.org/).

For comparison, Table 3-2 summarizes the approaches, including numbers of replicates, that form the bases of the performance characterization process recommended by SPADA/PSAA, PHAA/FSAPE, and as conducted by the BioWatch program (used during testing by LANL); Table 3-3 provides an additional comparison of approaches taken by SPADA and PHAA/FSAPE on several of the parameters discussed further in the report.

The SPADA process calls for testing 96 replicates to determine POD at AMDL; the PHAA/FSAPE process uses 9 replicates (3 runs by 3 operators) in a first phase to estimate the LOD followed by testing 60 replicates (20 runs by 3 operators) to determine LOD repeatability; and the BioWatch program tested 20 replicates followed by a further 20 if an unexpected result was obtained (e.g., a failure to detect). In its Defense Technology Objective CB.56 data transition packages, DOD assay developers commonly use several runs of 20 replicates for a total of 60 tests. A sample size of 60 likewise reflects Clinical and Laboratory Standards Institute (CLSI) guidance on limit of detection (CLSI 2004, 2012):

> As a reasonable compromise between precision and costs, a minimum number of 60 measurements (on both blank and low level samples) is suggested for establishing the LoD. This would typically be performed by the developer of the method. To verify a claimed LOD, use a minimum of 20 results at the claimed level and, if necessary, at the LoB [limit of blank]. This is usually performed by the laboratory or user. (CLSI 2004, section 4.2.1)

It should be noted that when using a method such as serial dilution to calculate assay LOD, the concentrations used should bracket the actual, obtained LOD in order to most correctly use curve fitting to calculate the value. Because of this, the committee notes that the two-step strategy employed by PHAA/FSAPE, which entails testing a limited number of replicates to initially estimate an approximate LOD, could be helpful in setting the appropriate concentration boundaries for the actual LOD calculation. It can also be noted that data obtained from lower concentrations of a serial dilution series help support confidence in results contained at higher concentrations (e.g., correctly detecting DNA in all tested samples at a lower concentration helps bolster confidence in the results from correctly detecting DNA in tested samples at a higher concentration). The committee has not attempted to place a statistical value on this point, however.

The relatively high numbers of replicates advocated by all of the approaches examined by the committee result in robust statistical power for this stage of the detection analysis. A statistical analysis of the SPADA and PHAA approaches undertaken by Signature Science on behalf of DHS

TABLE 3-2 Comparison of SPADA, PHAA/FSAPE, and BioWatch Acquisition Testing Approaches

	SPADA	
Characteristic	Approach	Testing Details
Detection Limit		
	Preestablished Acceptable Minimum Detection Limit (AMDL) of reference strain Two stages of testing: method developer stage and collaborative validation stage	Reference strain is spiked into matrix of interest Criteria require that false negative rate of the agent at the AMDL be less than 5%
	Method developer stage Determine probability of detection (POD) when reference strain is present at AMDL—goal is 0.95 POD at AMDL with 95% confidence	n = 96 replicates 0 or 1 nondetect = pass; >1 nondetect = fail If 95/96 samples, the 95% lower confidence limit on the POD is approx. 95%

FSAPE/PHAA (for environmental samples)[a]		**BioWatch Process** (LANL Gen-3 Testing)	
Approach	Testing Details	Approach	Testing Details
Determine analytical limit of detection (ALOD) and operational ALOD of reference strain	Multiple phases of testing to address these characteristics (Phases I–IV)	Determine probability of detection of reference strain spiked at different concentrations	Limit of detection determined to be concentration with 0.95 probability of detection
Phase I: Analytical Limit of Detection Test reference strain at concentrations of 0.025, 0.05, 0.1, 0.2, 0.4, 0.8 pg/reaction (rxn) In the absence of nontarget nucleic acid or environmental background	Estimate ALOD as lowest concentration that can be detected 95% of time n = 3 replicates by 3 operators (total = 9) for each concentration If 9/9 samples are able to detect the concentration, the 95% lower confidence limit on the ALOD is 66%	Step 1: Determine Probability of Detection Test reference strains at range of 8 or more concentrations of DNA (0 to 8192 copy numbers/rxn)	n = 20 replicates 0 unexpected results = pass; 1 nondetect, tested an additional 20 replicates; 39/40 detections = pass; >1 nondetect = fail If 20/20 samples are able to detect the concentration, the 95% lower confidence limit is 86%; for 39/40 samples the 95% lower confidence limit is 89%

continued

TABLE 3-2 Continued

	SPADA	
Characteristic	Approach	Testing Details
Detection Limit, *continued*		
	Collaborative validation stage 12 collaborators (4 at 3 sites) Each collaborator tests 12 samples of the reference strain; study must produce 10 valid data sets and report reproducibility and standard deviation (total = at least 120) See below for testing in absence of agent	Confirming POD in presence of agent Samples spiked at AMDL; matrix is dust-loaded collection medium (such as a BioWatch filter) POD at least 0.95 with 95% confidence limit

FSAPE/PHAA (for environmental samples)[a]		**BioWatch Process** (LANL Gen-3 Testing)	
Approach	Testing Details	Approach	Testing Details
Phase II: Intermediate precision/ repeatability Test reference strain spiked at the limit of detection calculated in Phase I Tested in absence of nontarget/ background	n = 20 replicates by 3 operators (total = 60) Criteria: If 0 negative results = pass For 60/60 samples, the 95% lower confidence limit of detection at the ALOD is 95% If ≥1 unexpected result, evaluate assay at next concentration above ALOD and/ or may modify the signatures or test (if so, assay must repeat process) 0-2 failures = "robust" if ALOD is 0.1 pg/rxn bacteria and 0.05 pg/rxn virus; 0-2 failures = "acceptable with limitation" if ALOD is 0.5 pg/rxn bacteria or 0.1 pg/ rxn virus; 3+ failures at 0.5 pg/rxn bacteria or 0.1 pg/rxn virus = "assay may not be acceptable for the intended application"	"Threshold of Probable Detection" measured at the BioWatch system level	2012-2014: System characterization and field tests conducted

continued

TABLE 3-2 Continued

Characteristic	SPADA	
	Approach	Testing Details
Detection Limit, *continued*		
	Not applicable	*Not applicable*
	Not applicable	*Not applicable*

FSAPE/PHAA (for environmental samples)[a]		**BioWatch Process** (LANL Gen-3 Testing)	
Approach	Testing Details	Approach	Testing Details
Phase III: Operational ALOD Test reference strain at concentrations of 0.025, 0.05, 0.1, 0.2, 0.4, 0.8 pg/rxn in presence of background (nontarget nucleic acid or environmental materials) Under FSAPE: Background of 20 "pooled processed BioWatch PCR-negative samples"	n = 3 replicates by 3 operators (total = 9) for each concentration Estimated operational analytical limit of detection calculated If 9/9 samples are able to detect the concentration, the 95% lower confidence limit on the measurement is 66%	*Not applicable*	*Not applicable*
Phase IV: Operational Intermediate precision/repeatability Test reference strain spiked at the operational ALOD calculated in Phase III in presence of background (nontarget nucleic acid or environmental materials)	n = 20 replicates by 3 operators (total = 60) Criteria: 0 false negatives expected If unexpected results, evaluate at next concentration above the ALOD (see description for Phase II) For 60/60 samples, the 95% lower confidence limit on the measurement is 95%	*Not applicable*	*Not applicable*

continued

TABLE 3-2 Continued

	SPADA	
Characteristic	Approach	Testing Details
Inclusivity Panel Testing		
	Test inclusivity panel organisms at concentration of AMDL	$n = 1$ 0 unexpected results = pass; If an unexpected result, re-test with $n = 96$, 0 failures allowed For 1/1 sample, the 95% lower confidence limit is 5%. For 95/96 samples, the 95% lower confidence limit of detection approx. 95%
Exclusivity Strain Panels		
	Test exclusivity panel organisms at concentration of 10× AMDL	$n = 1$ 0 unexpected results = pass; If an unexpected result, re-test with $n = 96$, 0 failures allowed For 95/96 samples, the 95% lower confidence limit of detection approx. 95%

FSAPE/PHAA (for environmental samples)[a]		**BioWatch Process** (LANL Gen-3 Testing)	
Approach	Testing Details	Approach	Testing Details
Demonstration of detectability Conducted prior to the limit of detection testing (Phases I-IV above) Inclusivity panel of strains are each tested at 1pg/rxn DNA or RNA (bacteria) and 0.5 pg/rxn (virus)	$n = 3$ ($n = 1$ run by 3 personnel when applicable) 0 unexpected results = pass; ≥1 unexpected result, "assay does not definitively meet performance standard" and will be investigated For 3/3 samples, the 95% lower confidence of detection approx. 37%	Step 2: Test using strains on SPADA inclusivity panels identified by SPADA panel chairs as high priority Step 3: Test using additional SPADA strains obtained by LANL	$n = 20$ replicates If 1 nondetect, tested an additional 20 replicates Tested at concentration of reference strain expected to give 0.995 POD based on the target strain probability of detection (Step 1) For vendor platform testing used concentration with 0.995 probability of detection for least sensitive system being evaluated
Phase V: Evaluation of Assay Exclusivity Test up to 10 organisms per pool at 100 pg/organism (total of 1000 pg/rxn) Tested in the absence of target	$n = 3$ ($n = 1$ run by 3 personnel when applicable) 0 unexpected results = pass; If 1-2 unexpected results, analyze each pool component (number of replicates unspecified) If all pool components individually test negative = pass If 3+ unexpected positive results = fail	Step 2: Test using strains on SPADA exclusivity panels identified by SPADA panel chairs as high priority Step 3: Test using additional SPADA strains obtained by LANL	$n = 20$ replicates If 1 unexpected result, tested an additional 20 replicates Tested at 10× the concentration of reference strain expected to give 0.995 probability of detection based on the target strain probability of detection (Step 1)

continued

TABLE 3-2 Continued

Characteristic	SPADA Approach	SPADA Testing Details
Exclusivity Strain Panels, *continued*		
	Collaborative validation stage 12 collaborators (4 at 3 sites) Each collaborator tests 12 samples; study must produce 10 valid data sets and report reproducibility and standard deviation (total = at least 120)	Confirming POD in absence of agent Samples spiked with a near neighbor at 10× AMDL; matrix is dust-loaded collection medium (such as a filter); 95% confidence limit that POD is 0.05 or less
Environmental Background Organisms and Substances		
	Testing pools of panel organisms at 10× AMDL each (up to 10 organisms per pool) Two types of testing: (a) in the presence of the target agent, and (b) in the absence of the target agent at AMDL	$n = 1$ 0 unexpected results = pass; If an unexpected result, re-test individual components with $n = 96$ For 1/1 sample, the 95% lower confidence limit is 5%; for 95/96 samples, the 95% lower confidence limit is approx. 95%

FSAPE/PHAA (for environmental samples)[a]		**BioWatch Process** (LANL Gen-3 Testing)	
Approach	Testing Details	Approach	Testing Details
Not applicable	*Not applicable*	*Not applicable*	*Not applicable*
Phase V: Evaluation of Assay Exclusivity Tested in the absence of target Up to 10 organisms per pool at 100 pg/organism (total of 1000 pg/rxn) Under FSAPE: Sample derived from known negative BioWatch samples (100; pooled 10/tube)	$n = 3$ 0 unexpected results = pass; If 1-2 unexpected results, analyze each component of the pool (number of replicates unspecified) all pool components individually test negative = pass 3+ unexpected positive results = fail	Step 4: Test using SPADA environmental panel organisms in pools Tested as individual organisms T1-11 and 8 pools	$n = 20$ replicates If an unexpected result, each strain from pool tested individually 20 times Tested at 10× the concentration of reference strain expected to give 0.995 POD based on the target strain probability of detection calculated in Step 1

TABLE 3-3 Comparison of Two Assay Performance Approaches on Several of the Parameters Discussed in the Report

Parameter	SPADA	PHAA/FSAPE
Limit of detection	Includes sufficient replicates for statistically robust determination of probability of detection at acceptable minimum detection level	Includes sufficient replicates for statistically robust determination of limit of detection and associated probability of detection at that quantity
Inclusivity strains	Strains represent a reasonable starting point but may not reflect sufficient genetic diversity for all pathogens, especially in light of new knowledge	Includes a larger number of strains than the SPADA panels; may call for testing more strains than required for sufficient BioWatch performance, particularly dependent on intended use of the assay
Exclusivity and environmental strains	Strains represent a reasonable starting point but may not reflect sufficient genetic diversity for all pathogens, especially in light of new knowledge	Includes a larger number of strains than the SPADA approach; may call for testing more strains than required for sufficient BioWatch performance, particularly dependent on intended use of the assay
Environmental substances	Includes a panel of environmental substances for interference testing; uses "dust-loaded filters" during reproducibility phases of collaborative validation	Does not specifically list environmental substances panel; some phases of FSAPE testing call for using pooled BioWatch filter samples to provide realistic environmental background
Process	Developed through stakeholder working groups; published as voluntary consensus standards in open literature through AOAC International	Developed as a government unique standard; experts within and outside federal government consulted but not published in open literature

reveals that, in both cases, the testing strategy "is highly protective against accepting the method if the true POD [probability of detection model at the AMDL, as defined by the SPADA standard] is lower than 0.95" (DHS 2013b, p. 9); similar analysis with regard to PHAA analytical LOD determination is found in DHS (2013a). For both standards approaches, this means that the true detection rate needs to be higher than 95 percent in order for the assay to have a good chance of passing the performance standard criteria. To meet the SPADA requirement of 95/96 or 96/96 detections with greater than 43 percent probability, the true detection rate must be over 98 percent; if the true detection rate is 95 percent the test will only have 4.4 percent probability of meeting the criteria (DHS 2013b). The initial phase of LOD estimation under PHAA guidelines (n = 9) has a lower statistical confidence associated with a 30 percent margin of error, but estimates undergo subsequent testing with a larger number of replicates (n = 60). To meet the PHAA requirement of 60/60 detections with greater that 50 percent probability, the true detection rate must be over 99 percent; if the true detection rate is 95 percent, the test will only have 4.6 percent probability of meeting the criteria (DHS 2013b). The assay LOD testing previously conducted as part of the BioNet study, initiated in 2005, similarly tested 60 replicates (using a strategy of 3 runs of 20 replicates tested on at least 2 days, by at least two operators, on at least two instruments). In this case, results of 58/60 or better were considered acceptable (Joan Gebhardt, Naval Medical Research Center, personal communication, January 6, 2015). PCR assays are highly sensitive, and validating the minimum LOD of the assay could be useful information because it helps set the maximum achievable performance. For a screening assay, this high level of statistical confidence in the probability of detection of representative inclusivity strains may be especially useful. As discussed above, a key consideration at a screening assay stage is avoiding false negative results.

Examining the methodology used by these various standards approaches and the statistical confidence presented in Table 3-1, the committee concluded that obtaining 60/60 detections is sufficient to infer a high degree of confidence that the assay will be able to detect DNA at that concentration with high probability of success. Even with acceptance criteria that would allow for two failures (58/60), the long-term probability of detection will still be close to at least 90 percent with 95 percent confidence. The committee concluded that this approach should be reasonable, when combined with appropriate data analysis and curve fitting, to determine an assay's LOD in the laboratory setting.

As noted previously, for decision makers the most relevant information is not necessarily the analytical laboratory LOD of an assay but whether the assay is sensitive enough to detect an amount of DNA reasonably expected to be present in a filter sample following an intentional release. Using the calculated laboratory LOD of the assay in concert with additional data and modeling from release to extracted sample would provide information to help determine whether the laboratory performance of the PCR assay meets such fit for purpose, operational requirements.

ASSAY SPECIFICITY: STRAIN AND SUBSTANCE PANELS

A BioWatch performance standard must sufficiently sample the known genetic diversity of a tested pathogen species such that a defined set of strains (the ones of biothreat concern) is detected, but closely related nonpathogenic species and strains are not detected. The standard also needs to provide a reasonable balance among competing goals: achieving maximal assay specificity, being feasible within the constraints of assay design and optimization, using available strain sequences and DNA against which to test the assay, and taking account of practical considerations such as time, cost, and who bears the primary responsibility for the burden of validation testing. As expected, there is an inherent tension over meeting these requirements.

An environmental pathogen detection assay for BioWatch should be designed and tested against three types of panels:

- Inclusivity: the set of pathogen strains the assay should detect.
- Exclusivity: the set of closely related species and strains the assay should not detect.
- Environmental background: the set of other organisms, including higher eukaryotes as well as additional microorganisms, that the assay should not detect through cross-reactivity. Potentially interfering background substances (e.g., dust, pollen) also should be included on defined environmental exclusivity panels.

The composition of the panels defined in assay standards are established through consultation with subject-matter experts knowledgeable about the genetic diversity of the pathogen and its near neighbors. Although strain panel composition varies among different standards, the general principles for determining which organisms and strains to include or exclude are similar. An example of the selection criteria used by one of the SPADA working groups in developing its inclusivity panel recommendations was provided by Luther Lindler (2014). The SPADA working group took account of the genetic relationships of this particular pathogen (including strains with pathogen-specific DNA sequences, strains with plasmid variations, and strains from various phylogenetic branches [Morelli et al. 2010]), as well as the virulence status of the strains (including strains with unusual genetic makeup known to be virulent in an animal model and strains with known virulence status verified by small-animal models). The selection of target inclusivity strains should generally seek to balance broad phylogenetic and geographic diversity within the realm of practical testing. The selection of exclusivity (near neighbor) and environmental organism panels generally should include a broad range of taxonomic diversity based upon available knowledge, data, and strain accessibility. The composition of the PHAA/FSAPE inclusivity and exclusivity panels was determined in coordination with subject experts at CDC (DHS 2014).

As Table 3-4 shows for a representative sample of biothreat agents, the inclusivity and exclusivity strain panels in the SPADA SMPRs generally contain fewer organisms than the inclusivity and exclusivity strain panels called for under the PHAA environmental sample approach or under FSAPE. As part of an informal interagency working group, Stephen Morse and Paul Jackson were asked to compare several of the SPADA and PHAA panels and provide their suggestions on whether the panels sufficiently sampled different species' genetic diversity. As can be seen from the table, representative pathogen species of interest vary significantly in their levels of genetic diversity, which is a key factor in the number of strains suggested for inclusion on the required panels by different approaches. The current SPADA panels should be reviewed in light of the new genetic knowledge available and the Morse-Jackson strain suggestions, given the time that has passed since most of the SPADA working groups established their recommendations (generally 2010-2011). Such a review would determine whether additions or alterations should be made to the panels.

In silico testing can be used to screen large numbers of inclusivity strains for predictions of assay detectability. The testing strategy should include *in silico* procedures (e.g., predictions of cross-reactivity based on sequence alignments undertaken during assay design and optimization) and laboratory analysis to validate the predicted performance. To keep the inclusivity panel size manageable for laboratory testing, a strategy for prioritizing the inclusivity panel strains is required. The committee suggests that inclusivity panel strains for laboratory characterization and validation of assay performance should sample the genetic diversity represented by available strain collections with the emphasis on prioritizing those strains that cause significant morbidity or mortality, have high transmissibility, or have a wide host range. The strain panel contents are likely to evolve over time as novel diversity is encountered, and it is likely that each agent of interest to BioWatch will have its own unique rate of such evolution.

For practical use in an environmental surveillance program such as BioWatch, the system also needs to function in the presence of a range of substances that could inhibit the assay detection (e.g., pollen, dirt). The SPADA SMPRs provide a list of environmental background substances against which assay performance should be tested to determine whether these substances inhibit or interfere with the results. The PHAA/FSAPE approach does not specifically identify a panel of background substances. At the request of DHS, LANL did not test assay analytical performance against the SPADA environmental substances panel, because system operational performance testing was subsequently undertaken by the program elsewhere. As a general principle, the committee agrees with the concept that assay analytical performance should be tested against environmental substances that might be present on filters in a realistic operational envi-

TABLE 3-4 Comparison of the Numbers of Inclusivity and Exclusivity Strains for a Representative Sample of Agents Under Different Standards Approaches

	Number of Strains and Substances	
Agent[a]	SPADA	LANL Assembly
A		
Inclusivity	15	13
Exclusivity	20	20
B		
Inclusivity	9	9
Exclusivity	10	7
C		
Inclusivity	16	15
Exclusivity	17	17
D and E		
Inclusivity	12	10
Inclusivity	24	24
Exclusivity	24	21
F		
Inclusivity	At least 2 from each major clade	*Ongoing*
Exclusivity	11	
Additional strains for testing	Based on bioinformatics screening	
Environmental panels (strains)	Other threat agents: 7 Bacteria in air/soil: 36 DNA viruses: 3 Microbial eukaryotes: 14 Higher eukaryotes: 18 Biol. insecticides: 5 *Total = 83*	Other threat agents: 3 of 4* Bacteria in air/soil: 33 DNA viruses: 3 Micro. eukaryotes: 8 Higher eukaryotes: 17 Biol. insecticides: 1 of 2
Environmental panels (substances)	32	N/A
Total strains and substances	262	Assembled 214 of 220 unique strains in SPADA panels

[a]Specific pathogen names have been replaced with a designation of "Agent A," "Agent B," etc. The full table and associated references are provided in Appendix A.
[b]These strain panels continue to evolve as new strains have been identified. Numbers listed reflect strains in DHS (2014).
[c]Based on an analysis undertaken by Paul Jackson and Stephen Morse (unpublished).

FSAPE / PHAA[b]	Jackson/ Morse	Jackson/Morse Suggestions[c]
48 61 (+ 7 on environmental panel)	15 22	Genetically monomorphic; SPADA panels sufficient for inclusivity and exclusivity
40 42	35 19	Genetically diverse; for inclusivity and exclusivity, supplement SPADA panels; provided suggestions for additional strains
46 37	40 32	Genetically diverse; for inclusivity, supplement SPADA panel; provided suggestions for additional strains; for exclusivity, use FSAPE panel
31 73 91	31 82 98	Genetically diverse; for inclusivity and exclusivity of D and E, use FSAPE panels
45 26		Did not review this agent
Arthropod: 20 Bacterial: 112 Botanical: 10 Fungal: 14 Protists: 10 Vertebrates: 20 Viral: 9 *Total = 195*	N/A	Did not comment on these panels
N/A	N/A	Did not comment on these panels
664		

ronment for informational purposes. If interference from environmental substances affects performance, modifications to sample preparation procedures or assay conditions may need to be made.

Strain access and quality control are fundamental issues for any performance standard. The work undertaken by LANL to assemble and quality control the inclusivity and exclusivity strains in published or forthcoming SPADA SMPRs for five pathogen species is illustrative of the challenges (AOAC 2011a,b,c, in preparation-a,-b; LANL 2011a). Not all strains of a pathogen that might be reported in the literature are easily accessible. Not all strain cultures or extracted DNA that LANL obtained matched canonical nucleic acid sequences for the strain; others failed quality control for other reasons. Although most strains could ultimately be obtained and verified, LANL was not able to obtain several of them, as Table 3-4 makes clear. The time, cost, and ability to obtain and verify strains would be expected to be greater the larger is the size of the required panels (such as for the more extensive PHAA/FSAPE panels).

As noted above, testing assay performance against inclusivity and exclusivity strains should be conducted through a combination of *in silico* prediction and laboratory analysis. *In silico* predictions of sequence alignment and potential detection or cross-reactivity rely on access to quality databases with as many accurate pathogen sequences as possible, but are an especially useful tool to screen large numbers of organisms quickly and economically and flag likely issues. Laboratory testing is more time-consuming, but is required to validate the expected performance.

Most standards approaches specify laboratory testing of inclusivity strains at the limit of detection (generally the quantity producing a 95 percent probability of detection, or a related modification such as the testing conducted by LANL at the concentration producing a 99.5 percent probability of detection). Exclusivity or environmental panel testing, on the other hand, is generally carried out at 10× the LOD or 10× the AMDL to maximize the chance of detecting an issue should one exist. These levels are appropriate and reflect general procedures in the community.

For exclusivity and environmental panel organisms, a strategy of sample pooling can be considered in order to increase testing efficiency. If a positive result is detected in a particular pooled sample, each agent in the pool then should be tested separately. Both SPADA and PHAA/FSAPE use a pooling strategy—SPADA for environmental organisms and PHAA/FSAPE for environmental organisms and exclusivity strains. Anecdotally, it is possible that assay sensitivity may decrease when pooling DNA from multiple organisms and the guidance documents reviewed by the committee generally limit pooling to a maximum of 10 organisms per pool. The reality that some pools may need to be deconvoluted and the strains tested individually should be kept in mind when planning validation time lines and budgets. Overall, however, the committee agrees that a strategy of sample pooling is reasonable in order to increase testing efficiency for exclusivity and environmental panels, rather than a requirement that each such organism be tested individually.

As with assay LOD, understanding the statistical levels of confidence around an assay's performance on inclusivity and exclusivity panel testing is important for interpreting the results. Different approaches use different testing strategies with different implications for statistical confidence. The SPADA SMPR testing for inclusivity, exclusivity, and environmental interference evaluates a single sample if the expected result is obtained. This provides only weak statistical confidence that the test will identify an issue should one exist. On the other hand, if an unexpected result is obtained, follow-up testing is done at high stringency (a further $n = 96$ samples). The PHAA/FSAPE approach for inclusivity on the full panel of inclusivity strains relies on an initial demonstration of detectability using three samples. The obtained threshold cycle (C_t) values are examined in comparison to the C_t value of the reference strain to judge how the assay performs across strains and whether there appear to be issues with the detectability of any of them. Exclusivity and environmental panel organisms also are only tested with $n = 3$ samples (three total runs of the panel strains by three different operators, one run per operator). The implication of both the SPADA and PHAA/FSAPE approaches for testing exclusivity and environmental strains is that the assays' potential false positive rate is not well characterized unless an issue is detected in the first tested sample and subsequently re-tested (under SPADA guidelines). For $n = 1/1$, there is 95 percent confidence that the lower confidence limit on the measurement is only 5 percent (e.g., the long-term detection rate is at least 5 percent); even if there is an assumed error rate of 20 percent, the sample would still have an 80 percent chance of passing the test. For $n = 3/3$, there will be 95 percent confidence that the long-term detection rate is at least 37 percent. With an assumed error rate of 20 percent, there will be a 51 percent rate of passing the test (DHS 2013a,b).

Under the SPADA approach, if the $n = 1$ sample does not produce the expected result of detection (inclusivity strains) or nondetection (exclusivity and background strains), a further 96 samples are tested. This produces stringent confidence around the results (should 95/96 detections be obtained based on the further testing, the assay will have a 95 percent probability of detecting the strain with a 95 percent lower confidence limit). On the other hand, this type of approach has the potential to lead to a significant increase in the number of tests required under the standard (e.g., were testing conducted on 20 strains and 10 produced failures in the $n = 1$ sample such that a further $n = 96$ replicates were required, an additional 960 samples would need to be performed).

High uncertainty around the false positive rate may be particularly problematic for a secondary assay, because assay specificity will be critical in ensuring that false positive results do not lead to high-consequence actions. The use of three or more independent signatures in a secondary assay (as is the case with the BioWatch assays through the LRN), and use of a decision algorithm that requires all of the signatures to be positive in order for the result to be considered a positive detection, mitigates the likelihood of a false positive in the overall detection algorithm, although the possibility of a false negative could exist if a signature fails. It appears to the committee that additional performance data at the inclusivity and, particularly, the exclusivity testing stage by testing a larger number of replicates than included in either SPADA or PHAA/FSAPE guidance would be valuable as part of a BioWatch performance standard. As discussed elsewhere in the chapter, obtaining and quality assuring strain DNA can represent a significant effort. Once this DNA has been assembled, conducting additional replicate testing in order to increase the statistical confidence associated with the results may not pose an extraordinary burden. In this regard, the committee notes that the LANL test and evaluation work conducted on behalf of BioWatch tested inclusivity and exclusivity strains using 20 replicates. If an unexpected result was obtained, 20 further replicates were tested. This yields a statistical lower confidence limit of 86 percent for 20/20 expected results and 89 percent for 39/40 samples.

A potential false positive or false negative performance rate on the assay of $1 - 0.86 = 0.14$ (14 percent) for a given strain may seem particularly problematic to program users and the committee considered this issue. Even for a secondary assay panel comprising three independent signatures, the potential false positive or false negative rate would be $0.14 \times 0.14 \times 0.14 = 0.0027$ or one incorrect result per 370 tests.

The committee noted that balancing the need to undertake a reasonable amount of laboratory assay testing (in both time and cost) with achieving high statistical certainty around assay performance is inherently challenging for a system such as BioWatch. The program has so far conducted over 7 million assays and must certainly keep its false positive rate low in order to be acceptable to jurisdictions. Its false negative rate must also be very low since a failure to detect an actual biothreat release would be potentially catastrophic. If one were to set the performance targets for laboratory validation testing of an assay at the sort of stringency level users may desire operationally (e.g., specifying that an assay false positive or false negative rate be less than 1 per 100,000 tests with 95 percent confidence, as a hypothetical example), it would not be possible or practical to conduct enough replicates to validate performance to this level. And furthermore, this process would reflect laboratory testing, rather than performance in the real-world environment of the jurisdictions.

The most feasible option to the committee is to conduct a manageable amount of laboratory testing at the assay performance characterization and validation stage, in order to gain confidence that the assay appears to perform reasonably. Ongoing analysis of assay performance data from verification in user laboratories and from operational deployment of the assay in the program will be critical to build a more complete understanding of assay performance over time. This ongoing data analysis will build confidence in the performance of the assay or reveal issues that need to be addressed.

As a result, the committee concluded that the approach and number of replicates used for inclusivity and exclusivity strain testing by LANL provides a reasonable level of confidence in assay performance for the laboratory validation stage. Combining the data obtained through this validation with ongoing performance monitoring should be considered in assessing confidence in the assay.

UNDERSTANDING ENVIRONMENTAL BACKGROUND

The BioWatch system is required to successfully detect a specific set of pathogenic organisms from the millions of nonpathogenic related or unrelated microorganisms present in the environment. This aim requires understanding how the assay result differs from a baseline. The program involves the collection of aerosol samples from multiple jurisdictions around the country and multiple locations within each jurisdiction. Because filter samples are collected daily, conditions such as the season, temperature, precipitation, presence of local disturbances such as construction, and other factors vary substantially over time for any given sample site. All of these factors make the background of microbes and substances out of which the pathogenic strains need to be detected extremely demanding.

How to detect a true signal from the noise of background is made more complicated still by the fact that the distribution of bacteria and viruses in different environments around the United States is not well characterized. Some BioWatch collectors are located in indoor, rather than outdoor, environments but indoor environmental background also is not well known.[4] Though culture was, for many years, the standard for characterization of microorganisms, many bacteria and viruses cannot be cultured. Advances in recent years in genomic sequencing have contributed substantially to sampling microbiomes in media such as water

---

[4] For example, the Alfred P. Sloan Foundation is currently supporting a research program to investigate "microbiology of the built environment" (see http://www.sloan.org/major-program-areas/basic-research/microbiology-of-the-built-environment/).

and soil samples, on surfaces, and in the air, yet it is common that a large fraction of the sequencing reads from such samples cannot be mapped to any known genome.

Investigations have been undertaken to explore the presence and composition of bacteria and near neighbors relevant to the current BioWatch program using sources such as archived filters from Environmental Protection Agency (EPA) air quality monitoring stations (Barns et al. 2005; Kuske et al. 2006; Merrill et al. 2006). Certain bacteria or near neighbors were detected more commonly than others, and notable variability was observed. These surveys also observed significant sequence identity between pathogens of interest and near neighbors, including near neighbors carrying partial or similar virulence genes, such as the pX01 plasmid found in *Bacillus anthracis* (Kuske et al. 2006).

It appears to the committee that public health users and others on the front line of decision making in BioWatch jurisdictions remain wary about the use of results in making decisions because of previous experiences with nonbiothreat environmental detections. This has been a bigger issue in some jurisdictions than others. For example, BioWatch jurisdiction "X" has experienced relatively high assay detections for tested pathogen "C," which has led the jurisdiction to stop using this particular assay (see Appendix A for further information). Specific investigations of these filters should occur and continue in order to uncover the root cause behind these results.

For BioWatch and its users, the practical implication is that to interpret assay data in a real-world situation, more accurate background information for existing collector locations is needed against which to compare results. Previously tested, threat agent–negative BioWatch filters from selected locations in the BioWatch jurisdictions could be used to undertake this environmental sampling, which could be carried out through deep metagenomic sequencing. Judicious sampling of microorganisms present in soil and water samples would also be informative. The significance of environmental background to confidence in assay results echoes a recommendation made in a previous NRC report (IOM and NRC 2011, pp. 9-10):

> RECOMMENDATION 5: As part of its response to the technical and operational challenges posed by the development and launch of Generation 3 BioWatch, DHS should collaborate with HHS, DoD, EPA, the National Science Foundation, and other agencies doing relevant work to develop and execute an aggressive research and development plan focused on (1) shorter-term goals to improve the capabilities and cost effectiveness of the environmental monitoring for airborne biological threats performed by the BioWatch system, and (2) longer-term goals to improve the knowledge base needed to support transformational inno-

vations in environmental biosurveillance.

Work in support of shorter term goals should focus on

- Advancement of the state of science needed for the development of an autonomous field-deployable detector with capabilities to meet Generation 3 BioWatch operational requirements and beyond.

Work supporting longer-term goals should include

- Temporal and spatial characterization of pathogen and near neighbor populations in air and natural reservoirs in urban areas, including those near BioWatch sites, and
- Participation in the work by others on host–pathogen interactions, surveillance, and epidemiologic research investigations and establishment of shared databases.

The use of existing resources such as used BioWatch filters to improve understanding of environmental background would build on the capacity of the BioWatch network as a tool. An improved understanding of background for that jurisdiction and location would help more broadly in assay data interpretation or in the prediction that such results might occur under certain sets of circumstances, and in understanding the context of the result. This understanding might help prevent a jurisdiction from dropping an assay in frustration.

In addition to helping put assay results in context, genomic sequence information obtained through further characterization of environmental background in BioWatch jurisdictions could aid PCR assay design (for example, in selection and *in silico* screening of target sequences, primers, and probes in an effort to make the assays as sensitive and specific as possible). It also is possible that results from further environmental background characterization could help guide selection of the most appropriate exclusivity and environmental strains with which to conduct the laboratory validation testing of BioWatch assays. As with selection of the inclusivity strains that should form part of the standard panels, employing a strategy to prioritize the most appropriate subset of environmental strains will be helpful in balancing the need to cover genetic diversity while reducing the size of strain panels and number of required tests.

Using Environmental Background to Understand Predictive Value of the Assay

Better understanding of environmental background would provide information on the expected probability of the tested organism or near neighbors in the specific jurisdictional environment. The predictive value of a test, particularly positive predictive value (PPV), is a concept that relates how likely it is that the condition is true when there is a positive test result ("true positives" divided by "positive test results"). The negative predictive value (NPV), on the other hand, reflects "true negatives" divided by "negative test results." These parameters are familiar from clinical and public health use. The determination of PPV is affected by prevalence. For BioWatch, the true prevalence of tested pathogens in local environments is difficult to determine, because little is known about the actual environmental background of the agents and how they vary geographically and temporally, and because the prevalence of the tested microorganisms is likely to be generally very low. For certain biothreat agents of potential relevance to BioWatch, such as Variola virus, the causative agent of smallpox, the environmental prevalence is expected to be zero. Other relevant biothreat agents are zoonotic and endemic, however, and may be found in certain U.S. environments under certain conditions. As a result, the prevalence is not necessarily too low to be relevant in all cases. Earlier in the program's history, for example, testing included a species of bacteria that infects livestock animals such as cattle, pigs, and sheep, and wildlife such as bison; this organism was subsequently dropped from BioWatch because of the frequency of environmental detections. Decision making by state and local jurisdictions would be strengthened by a better understanding of assays' PPV in the local context, could this be obtained. The committee notes that the newly released SPADA SMPR for Variola virus (AOAC 2014) asks for the detection assay's PPV to be reported given that the virus has been eliminated from the environment.

MICROBIAL DIVERSITY

Microorganisms are diverse. Bacteria, for example, may undergo mutations during replication, exchange genetic information, experience deletions or rearrangements in their genomes, or gain and lose plasmids (additional, extra-chromosomal pieces of DNA) associated with virulence. Viruses, especially RNA viruses because of an error-prone polymerase (Smidansky et al. 2008), may also mutate, or viruses with segmented genomes, such as influenza, undergo recombination (Bean et al. 1980). In addition, viruses exhibit a broad inherent diversity in mammals that far exceeds the known diversity in humans (Anthony et al. 2013), and novel human viruses are being continually discovered (Chiu 2013). This diversity is of direct relevance to BioWatch because nearly all of the microbial agents targeted by the program are zoonotic in origin. Appendix A provides an additional figure that attempts to capture measures of variation involved with representative pathogenic agents.

Several concepts have been proposed to express the range of microbial diversity. Viruses may evolve through mutation into distinct phylogenetic groups, or "clades," resulting in independent, potentially highly diverse strains circulating in the population. Even within an infected individual, there can be viral "quasi species" which reflect a distribution of closely related viral genomes in a population. It is thus critical that PCR assays targeting these viruses use primers derived from highly conserved regions of the genome to be able to capture the diversity of viruses within individuals as well as populations. In many cases, a single viral reference sequence may not be sufficient to detect all viral genomes within a population (Domingo et al. 2012), and the availability of multiple reference sequences spanning the phylogenetic "space" may be needed. The viral reference database for assay design also needs to be regularly updated given the high mutation rate for RNA viruses. A recent genomic study of Ebola viruses from the 2014 West African outbreak found that the 2014 genomes differed from the PCR primers and probes for four separate assays that had been deployed for Ebola and pan-filovirus diagnostics (Gire et al. 2014).

For organisms such as bacteria, defining the pan-genome is an effort to represent the species diversity that exists. The pan-genome represents the full set of genes present in all of the strains of a species. Thus, the known pan-genome is likely to grow as new strains are identified and sequenced.[5] The pan-genome includes a core genome, containing the absolutely conserved genes present in all members of the species, and an accessory genome, representing the elements that are present in only one or a subset of strains. Whereas the core genome of a species is associated with many of its fundamental characteristics, it has been suggested that the accessory genome is frequently associated with traits of interest such as antibiotic resistance and virulence (Tettelin et al. 2005). As new strains are sequenced the accessory genome may grow while the core genome is reduced.

An example from the bacterial species *Burkholderia pseudomallei* illustrates the challenges that can be presented for PCR assay development. To avoid a failure to detect the particular species of interest (a false negative), assays target DNA signatures drawn from the core genome of the species. To avoid detecting a near neighbor that is not the species of interest (a false positive result), assays target those core genome sequences that are not shared with the other species. As can be seen in Figure 3-3, the *Burkholderia* genus exhibits very high diversity and there is only a small portion of the overall *B. pseudomallei* genome that is suitable for use in designing an effective PCR assay. As more and more genomic informa-

---

[5] Alternatively, as genetic sequence information increases, a pathogen previously identified as a strain of a particular species could be reclassified as belonging to a different species. In that case, its genetic information would no longer be part of the pan-genome for the initial species.

A

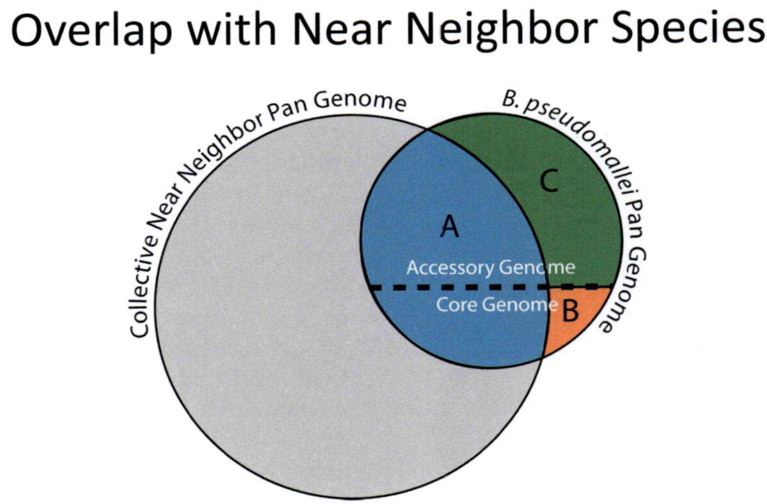

Characterization of *Burkholderia pseudomallei* using unique genomic components

B

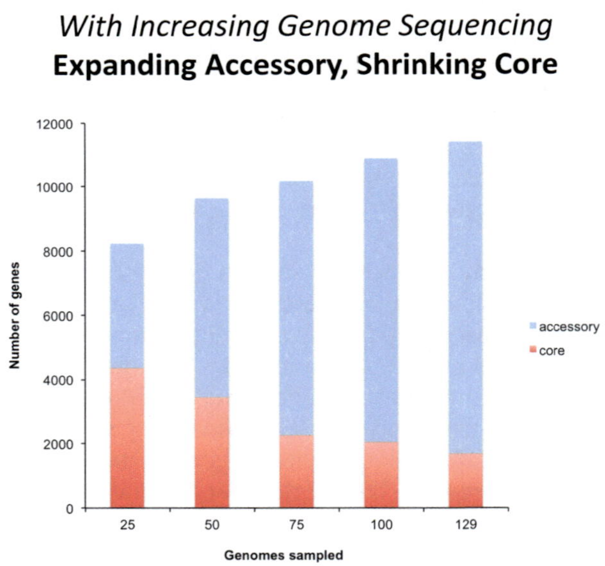

FIGURE 3-3 A. Representation of the *Burkholderia pseudomallei* pan-genome. **B.** Effect this has on the selection of DNA signatures that can be used in designing PCR assays.
SOURCE: J. Sahl, P. Keim, and D. Wagner, unpublished data, used with permission.

tion has become available on *B. pseudomallei* strains and on closely related *Burkholderia* species, what is recognized as the conserved, core genome has shrunken and the accessory genome has grown. This presents a huge challenge to identifying DNA signatures that can be distinctive for particular subsets of *B. pseudomallei*. Furthermore, as knowledge of the *Burkholderia* genus has expanded, it is being recognized that pathogens *B. pseudomallei* and *B. mallei* may need to be treated as a single assay target rather than as two separate targets, because of their genetic similarity. Despite causing vastly different disease manifestations, there are now no uniquely identifying signatures available for *B. mallei* (Keim 2014).

Even for bacteria that exhibit much less genetic diversity than the *Burkholderia* genus, a given bacterial strain can vary unexpectedly. For example, investigations into the *Bacillus anthracis* Ames strain used in the 2001 anthrax mailings identified morphological differences among colonies within the Ames strain and associated genetic changes, including insertions, deletions, or single nucleotide alterations in the chromosomal DNA or in the plasmid pX01 (NRC 2011; Rasko et al. 2011). Should a PCR assay's primers and probe target sequences exhibit such variations, binding and amplification could potentially be affected.

To help deal with the challenges posed by genetic diversity, it may in some cases be useful to incorporate into the BioWatch program the use of one or more additional near-neighbor rule-out assays as part of the decision-making algorithm (e.g., the development and use of PCR-based or other assays, beyond the existing screening and secondary assays, that would be designed to detect expected nonpathogen near neighbors). These could be incorporated into the program for pathogens that exhibit high genetic variability that makes it challenging to design and validate screening and secondary assays against a limited number of target DNA sequences with sufficient powers of discrimination. This option might be considered for use with certain pathogens in the context of certain collector sites or certain jurisdictions where it has been challenging to obtain PCR assays with sufficient specificity. This situation also suggests that a single set of initial and secondary assays sufficient for all BioWatch jurisdictions and all collection sites may not be feasible.

ASSAY PRECISION AND OTHER PERFORMANCE PARAMETERS

Assay performance characterization undertaken to validate performance standards needs to consider several additional factors such as precision and robustness. Assay precision represents the closeness of the values of results obtained through independent tests of a sample. It can include two dimensions—repeatability (measurements carried out under identical conditions and over a short period of time) and reproducibility (measurements carried out using different reagent lots, different operators, and over different days).

In addition, the accuracy of the assay can be determined in situations where results from the tested assay or method can be compared with results from a standard reference method or with a true value known by other means. For analytical testing of a real-time PCR assay starting from known quantities of a purified nucleic acid, the accuracy of the assay result can at least be compared with the starting reference value.[6] As the BioWatch program validates new assays or new detection technologies, it also can compare the performance of the new candidate method to the existing assay system.

POSITIVE AND NEGATIVE PCR ASSAY CONTROLS

Appropriate positive and negative controls are required when conducting all real-time PCR assays to ensure that the reagents and instruments are operating correctly, that there is no detectable contamination, and that PCR inhibitors are not present. These controls should be specified in any performance standard. Rather than compare and contrast descriptions of assay controls used by LANL in its extensive testing or discussed in existing performance standards, the committee provides the principle types of controls it judges are necessary for inclusion in a performance standard for a real-time pathogen detection assay to meet BioWatch's needs. These are

Positive controls

- Positive target amplification control (external positive control): contains all of the reaction components along with reference DNA that should be amplified by the assay; used to help rule out amplification failure if a negative sample result is obtained.
- Positive amplification control (inhibition control): contains materials designed to amplify a known, different region of DNA (such as the bacterial 16S gene) along with a test sample; used to help rule out an amplification failure due to the presence of substances that inhibit PCR in the test samples. In addition to running the inhibitor control within the background of a test sample, the inhibitor control should be run by itself to rule out a failure of amplification due to a failure of the 16S assay or reagents.

[6] Note that when creating test samples using dilutions series the actual concentration of DNA in a given sample will be subject to a statistical distribution. The effects of a Poisson distribution of genome copies will be more noticeable at very low copy numbers.

Negative controls

- No template control: contains all of the reaction components in the absence of target DNA; used to detect contamination or some nonspecific amplification products.

Instrument and operator performance controls

- Quality and uniformity controls: known samples run regularly that help the laboratory assess that procedures and instruments are performing as expected. LANL's testing and evaluation, for example, included the following:
  - — Daily quality control samples that verified the concentration of DNA aliquots used in testing and correct preparation of dilution series; and
  - — Monthly uniformity samples that contained the same reaction in all wells of a 96-well assay plate, to verify that similar results were obtained across the plate.
- Prior to validation testing, the assay or instrument developer should provide information on the expected acceptable performance of the assay on the instrument or platform, such as specified ranges for C_t values. This type of information is used to ensure, for example, that the results produced by positive amplification controls are in the correct range.

Sufficient positive and negative controls should be included to ensure that the PCR process was performed properly and to reliably indicate when there has been some type of mitigating result. Quality control and proficiency test samples also should be run as part of a comprehensive quality assurance (QA) framework. Ongoing monitoring and evaluation of assay and operational performance through a QA program is discussed further in Chapter 4.

TESTING LABORATORY SETUP AND MANAGEMENT

Assay performance and validation testing should be conducted using a laboratory setup designed to minimize the potential for contamination and with systems to record and manage the data obtained. International standards such as ISO 17025 provide guidance for testing laboratories and should be consulted.

The report does not delve into appropriate laboratory setup in detail, but points, for example, to the procedures implemented at LANL for the assay and platform testing it conducted on behalf of the BioWatch program (LANL 2010). Different stages of the assay testing process were conducted in separate rooms with directional sample flow from one stage of analysis to another (see Figure 3-4).[7] Weekly swipe testing also was conducted to monitor for cross-contaminating DNA. The time, space, and cost to set up and maintain the appropriate type of laboratory operation (with associated, trained personnel) is not insignificant.

REFERENCE MATERIAL AVAILABILITY

A critical dimension of assay characterization and validation, regardless of the performance standard, is the use of quality-controlled reference materials. Reference materials must be available to BioWatch assay developers and laboratories conducting validation in quantities sufficient for the required testing.

For BioWatch assay development and testing, access is required to at least extracted nucleic acids from inclusivity, exclusivity, and environmental background organisms. For ongoing or large-scale testing, having access to microbial cultures from which new batches of DNA can be extracted and purified is helpful. The BioWatch program has also conducted some system testing using live microbial agents. The committee recognizes that access to certain live biothreat agents may be limited to designated parties or locations that are licensed to handle these organisms under appropriate safety and security regulations. Even with select agent registration it can reportedly be difficult to access select agent materials and, in particular, federal agencies may not be willing to distribute select agents to nonfederal entities. Because of specific restrictions on the possession of, for example, Variola virus cell culture stocks and DNA, special considerations arise when undertaking assay performance testing and validation for orthopox detection assays.

To reduce the need for high-level biosafety facilities and to clarify access, establishing mechanisms to provide assay developers and validation testing laboratories with necessary testing materials would be helpful. Non–live agent strain materials that could be used for conducting PCR assay testing include sterilized or killed agents, purified genomic DNA, or synthetic or cloned DNA fragments. To address orthopox assay testing needs, for example, the molecular genetics and assay development communities are currently wrestling with the issue of whether candidate gene fragments can be used as an acceptable substitute to

[7] Preparation of PCR master mix and preparation and aliquoting of DNA are performed in the same room, although the steps are separated on either side of the room. Note that for clinical and public health PCR testing, spatial segregation between these areas is generally recommended. Thus this specific setup may not be the one in operational use in BioWatch laboratories.

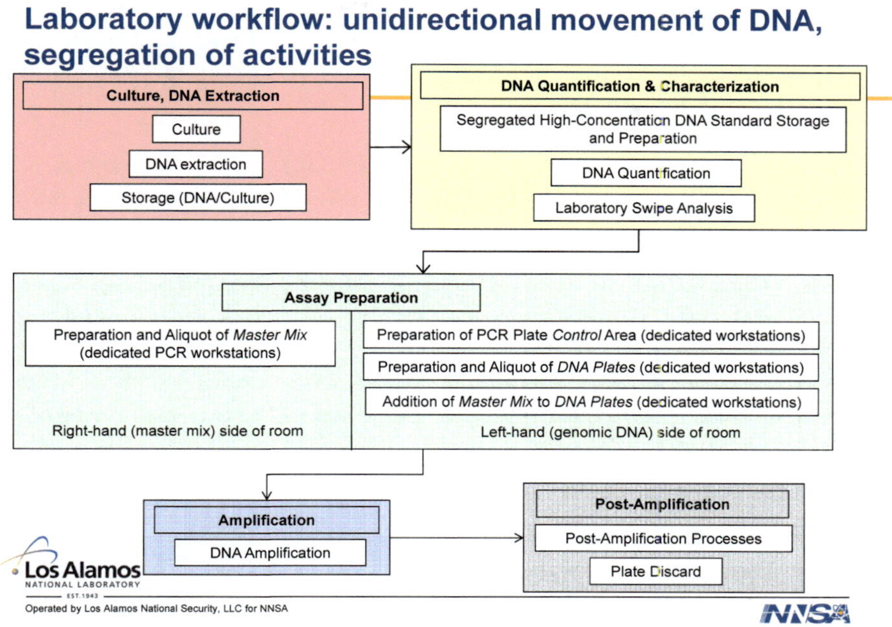

FIGURE 3-4 Laboratory organization used by LANL in conducting assay testing.
SOURCE: Los Alamos National Laboratory, used with permission.

genomic DNA or live virus, or whether all Variola materials and testing would need to be conducted at the CDC's Variola laboratory, which could place a burden on the CDC's Biosecurity Level 4 resources. The committee supports efforts to address the issue of which types of strain reference materials are necessary for biodetection assay testing and how they can be provided, and notes, as well, that the need for access to relevant reference materials does not imply that all such materials should be held in a single repository.

Awareness of who holds nucleic acids from which microorganisms and how to obtain these materials is an important issue for the biodetection community and has the potential to limit assay development and validation efforts. A clear list of who maintains which strains and strain nucleic acids across federal agencies and through other known sources currently is not available. The committee's understanding is that discussions were previously held among federal agencies and relevant participants on creating such a catalogue, informational reference, or virtual database, but the effort did not come to fruition.

There are several sources from which relevant strain cultures and extracted strain nucleic acids can be obtained. Acquisition and subsequent use of materials from all sources is likely to be governed by applicable material transfer agreements:

- *Los Alamos National Laboratory*: In 2009, DHS contracted with LANL to assemble the inclusivity and exclusivity strain panels recommended by the SPADA SMPRs that address aerosol collection of representative biothreat agents. The materials that LANL assembled, particularly culture stocks from which new batches of DNA could be produced and purified, remain available for further assay or device testing.
- *Critical Reagents Program*: As discussed in Chapter 2, the CRP provides a source for reference materials to DOD as well as to other agency purchasers.
- *American Type Culture Collection (ATCC)*: ATCC is a nonprofit center that maintains and sells microorganisms, cells, and other biological materials (non–select agents).
- *BEI Resources (BEI)*: BEI was initially established by the National Institute for Allergy and Infectious Diseases of the National Institutes of Health to support infectious disease research, including maintenance and provision of microbiological research materials. It is currently managed under a contract to ATCC.
- *Other Agency Collections:* DHS and CDC have identified strains listed on the PHAA panels for agent detection assays as part of CDC LRN assay validation testing and documentation.
- *Other Resources*:
  - Individual researchers may be sources for particular strains. For example, Paul Keim is recognized as having assembled a substantial collection of *Burkholderia* species in the course of his research on that genus.
  - Private companies may hold relevant strains from their research and development of different types of detection devices. For example, MRI Global has served as the independent testing laboratory to conduct validation following AOAC and SPADA requirements and maintains ISO 17025 accreditation under the American Association for Laboratory Accreditation. It also formerly served as an "Accredited Reference Material Producer" to the ISO 34 standard.[8]

[8] MRI Global recently canceled its ISO 34 certification because it did not have active business in this particular area (Chris Bailey, MRI Global, personal communication, January 5, 2015).

The initial screening assay and associated reagents used by BioWatch are provided to the jurisdictional laboratories through the DOD CRP. At the time that LANL assembled SPADA panel strains in order to conduct performance testing, it obtained some materials from CRP, some from repositories such as BEI and ATCC, and some from private collections and other sources. In parallel, the CRP undertook an effort to obtain stocks for additional microorganisms that it did not previously hold. As a result, the CRP now strives to hold as reference materials the strains listed on SPADA inclusivity and exclusivity panels that are biological weapons agents, and the strains held by CRP are in the process of being sequenced (Bruce Goodwin, CRP, personal communication, January 12, 2015).[9] The resultant data raise the question for the microbiology, molecular genetics, and population genetics communities about what degree of sequence similarity is sufficient for something to be considered the same strain and to proceed with using it for the panels, given the potential for mutations and plasmid changes over time. The CRP is wrestling with this issue (Goodwin 2014; Omberg 2014).

The secondary assay and associated reagents used by BioWatch are provided to the jurisdictional laboratories through the CDC LRN. The CDC has adopted the PHAA performance standard, which includes testing procedures and strain panels for clinical samples and procedures and strain panels for environmental samples. The CDC has been submitting LRN assays to the Food and Drug Administration (FDA) for 510(k) approval (clinical diagnostic component); it has so far completed this submission for one LRN agent detection assay and has others in the process (Toby Merlin and Harvey Holmes, CDC, personal communication, December 2, 2014). As a result, sources for the relevant pathogen strains on PHAA panels for certain agents have been identified, and strain material has been extracted. In this case, nucleic acids from the identified strains are requested through the DHS Science and Technology (S&T) Directorate, which communicates with the experts who possess and maintain the strains in their laboratories. The committee does not have detailed information on the quality control processes used in verifying the panel strain materials, but these experts provide certificates of analysis on which CDC relies for strain authentication. As noted, the subject experts continue to maintain the strains in their own laboratories and provide them on request. However, these experts are not professional repositories and a streamlined mechanism or repository to make standard materials from the PHAA panels available to the broader community does not cur-

[9] The CRP indicated that it focuses on strains that are biological weapons agents (BWAs) and is generally not in the practice of producing reference materials for non-BWAs as these materials may be available through other entities.

rently exist and remains a limitation (Toby Merlin and Harvey Holmes, CDC, personal communication, December 2, 2014). It might be feasible for panel materials eventually to be transferred to another entity to serve as the resource to provide them to DHS or others, as necessary for future assay performance and validation testing, but this option would need to be explored.

Quality Control

LANL was able to obtain the majority of the SPADA strains it sought—214 out of a total of 220 strains in SPADA inclusivity, exclusivity, and environmental strain panels were obtained by time of its report to DHS in 2011 (LANL 2011a). Some of these strains were obtained as cultures and others as purified nucleic acids.

LANL scientists went through an extensive process to obtain and qualify these 214 strains to verify that the materials were the agents specified and to assess the purity, quality, and concentration of the nucleic acid prior to use (Omberg 2014). The multistep characterization varied with species based on published information, but

> involved a combination of PCR for plasmid identity and/or specific genomic signatures, sequencing of the 16S rDNA gene, multi-locus sequence typing (MLST), canonical single nucleotide polymorphism (SNP) analysis, amplified fragment length polymorphism (AFLP) analysis, and variable number tandem repeat (VNTR) region analysis to confirm strain identity. Samples were characterized at both the species level and the strain level. (LANL 2011a, p. 22)

A table summarizing the strain characterization procedures used by LANL for different species is included in Appendix A.

LANL investigators reported several challenges. It was difficult to identify strain sources for certain samples of DNA and, in some cases, multiple or inconsistent nomenclature in the literature complicated the ability to determine that they were obtaining the correct strain. Once obtained, some strains were discovered to be contaminated or to have lost canonical sequences (e.g., exhibited mutations in the set of single nucleotide polymorphisms used to type the strain or had lost particular plasmids normally contained in that strain of that particular pathogen). LANL reportedly went through three different sources to obtain one particular strain. It also noted that some species and strains in the SPADA panels were very precisely specified whereas others, particularly on the environmental panels, were specified only generally (e.g., pine pollen). LANL did not undertake the same level of extensive characterization of the SPADA environmental panel materials as it undertook for the inclusivity and exclusivity panels because there was insufficient literature on how to proceed for many of the organisms.

LANL's work assembling and qualifying strain reference materials for subsequent assay performance evaluation has implications for a BioWatch performance standard and the qualified reference materials that are needed to support the standard. It was not a trivial exercise or inexpensive process for LANL to identify, assemble, and quality control the materials they obtained, which represented the majority (but not entirety) of the inclusivity and exclusivity strains recommended by SPADA for the representative set of agents they examined. Similarly, MRI Global assembled SPADA strain panels (inclusivity targets, near-neighbor and background exclusivity panels, and potentially interfering environmental substances) for at least one biothreat agent of relevance to BioWatch. The effort in time and costs to identify sources for these strains or strain nucleic acids, acquire them, and undertake quality control and sample preparation were similarly nontrivial, although the committee does not have a precise figure for these costs (Matthew Davenport, Johns Hopkins University Applied Physics Laboratory, personal communication, January 8, 2015). An important question posed by LANL and others is "how good is enough" for ensuring the microbial strain identity and quality of DNA in the panels that will be used for assay validation. Both the number of required strains and the level to which they need to be qualified will affect the cost required to assemble and maintain reference material collections. The strain panels recommended by PHAA and FSAPE are more extensive in number than the SPADA panels, which may make this process even more complicated.

TIME AND COST CONSIDERATIONS

There are obvious tradeoffs between how precisely reference materials can be qualified, how thoroughly the performance of an assay can be characterized and validated, how fast this can be accomplished, and how much the effort will cost:

- The greater the number of inclusivity, exclusivity, and environmental materials against which assays should be tested, the longer it will take to obtain and validate these materials. The largest investment of time and cost is incurred during initial panel assembly, although there are ongoing production and quality control expenses associated with providing the materials to users (making additional culture stocks or extracted DNA as needed, ongoing quality control monitoring of existing lots, and testing of new lots, etc.).

- The more precisely assay reagents and reference materials need to be quality-tested initially and the more frequently they need to be re-tested, the more the time and cost will increase. For example, some materials may be subject to higher mutation rates or to the frequency with which plasmid gain or loss occurs. DHS should explore the question of "what's close enough" with regard to genetic changes for DNA to be considered the same strain, in order to inform its assay and system testing.
- The better the available data curation—for example, the existence of high-quality sequence data deposited into reference databases and documentation on other strain characterization parameters—the greater is the likelihood that some labor and resource costs may be reduced.
- The greater the number of inclusivity, exclusivity, and environmental materials against which assays should be tested, the more precisely assay specificity will be known but the higher the time and cost to conduct the assay testing. Similarly, requiring testing with a greater number of replicates will increase statistical confidence in the assay performance results but increase time and cost.

Table 3-5 summarizes the cost information provided to the committee from several sources. The cost estimates for performance testing given in the table include the stages of laboratory testing undertaken to determine assay probability of detection and LOD, assay specificity using inclusivity, exclusivity, and environmental panels, and, where applicable, testing using BioWatch filters to provide background. The costs do not include multisite collaborative validation, a component that is found in the AOAC validation guidance documents. In addition to the information in Table 3-5, a commercial vendor has taken a PCR detection system through a SPADA-based validation process. It took BioFire 2-3 years and roughly $3+ million dollars to validate its RAZOR EX device to the full AOAC/SPADA standard for the detection of *B. anthracis*, which included performance testing conducted by BioFire, production of testing kits for additional testing, program management, and report writing, but did not include the independent laboratory testing conducted by MRI Global to address AOAC/SPADA requirements for multilaboratory validation (Matthew Scullion, BioFire, personal communication, December 12, 2014). Undertaking independent laboratory validation at MRI Global to repeat the vendor testing conducted by BioFire took roughly a month and cost approximately $300,000, although it is estimated that costs for independent validation under the SPADA approach might be able to be reduced to something on the order of $150,000 based on lessons learned and/or revisions in the required testing (Matthew Davenport, personal communi-

TABLE 3-5 Estimated Time and Cost Associated with Assay Testing and Reference Material Provision

Step	**Details**	**Completion Time** (approx.)	**Cost** (approx.)
Assembly and quality control of reference materials (strain panels)	SPADA (214 strains obtained by LANL)	18 months	$1. 2M[a]
	PHAA (to identify strain sources and develop extracted material)	2 years	$0.5-1M
Testing laboratory setup costs	LANL for BioWatch assay testing		$2.1M
Assay performance testing	BioWatch process	7 months	$0.6M
	SPADA/PSAA	6 months	$0.4M
	PHAA	2 or more years	$1-2M[b]
Platform performance testing (comparison of two Gen-3 vendor systems with reference system)	LANL for BioWatch vendor system testing	9 months, setup to completion	$4.4M
Ongoing assay and reference material provision	PCR assays and reagents; most strains on SPADA panels that are bioweapons agents	Available through CRP catalogue	$909 per strain vial (~200 ng DNA/vial). Strains sold individually (not in panels)
	PCR Mastermix		$4 per reaction (sold in vials suitable for variable numbers of runs and using variable technologies)
	PCR positive plasmid controls		$90 per 50-μl tube
	PHAA panels and LRN assays	Coordinated through DHS S&T Directorate	Available through subject-matter experts who hold the identified strains in their laboratories

[a]Cost includes nucleic acid acquisition and quality control. The unburdened nucleic acid acquisition cost was reportedly $605k (Omberg 2014); the cost to acquire nucleic acid including labor to extract DNA was reportedly $1.4M (Walter 2014b). The differences reported

continued

TABLE 3-5 Continued

in quality control costs ($560k in Omberg [2014] and $982k in Walter [2014b]) presumably represent differences in how these figures are calculated and what costs are included.
[b]A figure of 2 or more years and $1.2M per agent was reported by Walter (2014b). A figure of $1.8M "end to end" over 2-3 years per agent was reported by scientists at CDC, who noted that this figure reflects elements associated with LRN assay submission under FDA regulations. The CDC scientists suggested that the figure might be reduced by roughly half for environmental assay validation that did not involve FDA approval (Toby Merlin and Harvey Holmes, CDC, personal communication, December 2, 2014).
SOURCES: LANL 2010, 2011a,b, 2012a,b; CRP 2015; Walter 2014b; Toby Merlin and Harvey Holmes, CDC, personal communication.

cation, January 8, 2015). BioFire's and MRI Global's costs for this activity were supported primarily by DHS.[10]

These expenses are not insignificant. On the other hand, investments have already been made and are continuing to be made in assembling relevant panels of strains and establishing workflow processes in independent testing laboratories, both at private companies and in national laboratories. Initial work to assemble SPADA panel strains has been undertaken by CRP and LANL, while work identifying sources from which to assemble PHAA/FSAPE strains has occurred and appears to be ongoing by DHS and CDC. These efforts provide a base for undertaking future performance characterization and validation studies should the need arise.

An important consideration is who should bear the burden in time and cost of which parts of assay development, performance testing, and validation. For clinical diagnostic use, a significant burden is on the medical assay or device developer to sufficiently demonstrate performance in order to obtain FDA approval. The health care market is a key driver of advancing technology, and the potential financial rewards to developers for a successful product help support this system. Although an overseeing regulatory authority such as the FDA does not exist in the environmental biosurveillance realm, biodefense remains a potentially significant market. Government agencies serve as substantial buyers of technology; however, different local, state, and federal agencies and programs may need different types of assays and devices to meet different mission needs.

[10] The RAZOR EX device was developed as a portable field detection system using pouches of reagents; it is not the same as the system of collectors and laboratory-based PCR assays used in the BioWatch program. However, it provides the only data point available to the committee on the time and cost associated with validating a biothreat detection device to the full AOAC/SPADA standard.

One approach that could be considered would be for assay and device developers to be required to undertake initial studies to validate assay specificity using extensive *in silico* screening as well as a reasonable amount of laboratory testing against a set of inclusivity, exclusivity, and environmental organisms and substances. For example, developers might be asked to document performance testing against some or all of the SPADA strains. Further validation conducted by DHS through independent laboratories could be used to confirm the performance reported by the developer against a subset of the same strains and to conduct performance validation testing against additional strains to further sample genetic diversity (e.g., using additional priority strains drawn from the PHAA/FSAPE panels). Such a model would enable validation costs to be shared and would mean that validation testing is undertaken at several steps in the process from assay design to operational use. Both SPADA and PHAA/FSAPE guidelines list potential tools and databases that can be used for *in silico* screening (DHS 2011, 2014; AOAC 2014).

In estimating the time lines and budgets necessary for assay performance characterization and validation, it will also be important to take into account the implications of obtaining "failures" at different points in testing. Both SPADA and PHAA/FSAPE pool DNA from exclusivity and/or environmental strains to reduce test numbers. But for each pool that produces an unexpected assay result, individual agents will still need to be tested separately and so the potential cost savings from a pooling strategy, for example, will not be perfect.

USE OF PRIOR AND ONGOING PERFORMANCE TEST INFORMATION

CDC is currently testing its LRN assays, which are used by BioWatch as secondary assays, to the PHAA standard for submission to FDA (clinical diagnostic use). PHAA/FSAPE guidance includes more extensive inclusivity and exclusivity strain panels than SPADA. PHAA/FSAPE also uses a fairly similar overall testing approach to that suggested by the committee, although the committee has suggested conducting a larger number of exclusivity and environmental panel replicates than the $n = 3$ specified under PHAA/FSAPE. If CDC is willing to share its performance testing data with the BioWatch program, the committee does not mean to imply that the BioWatch program should duplicate all of the time, effort, and cost that will have gone into CDC's work on these assays.

LANL, on behalf of DHS, has conducted testing on several of the CRP assays, which are used by BioWatch as screening assays. The approach used by LANL involved SPADA panels and was not exactly the same as the approach suggested by the committee, but is also very similar. The committee likewise does not intend to imply that the BioWatch program should redo the existing foundation of work that has been conducted or is ongoing. The committee wishes to emphasize three points, however:

- The need to annually revisit all strain panels in light of new genetic knowledge and, for example, specific suggestions such as those made by Paul Jackson and Stephen Morse with regard to SPADA panels (see Appendix A for further details on these suggestions).
- The need for BioWatch to better understand actual environmental background to inform assay design (target sequences, primers, and probes), assay testing (identification of the most important inclusivity, exclusivity, and environmental strains and substances), and user understanding of assay performance. The committee stresses the usefulness of this knowledge, which remains very limited.
- The need for communication with laboratory experts and jurisdictional officials on the detailed performance data obtained, the necessary limits to what can be accomplished through *in silico* and laboratory assay testing, and how the data can be interpreted and used. The committee aims to recognize the valuable efforts that have been made to obtain assay performance data but does not wish the message to be that DHS, CDC, and others have conducted all of the necessary testing and that jurisdictions should trust that fact without enabling data access and discussions.

In this chapter, the committee explored key factors that will affect whether a PCR assay standard is suitable for performance testing and validation as part of the BioWatch program. However, the steps that surround the PCR play critical and interconnected roles in the overall system. Supporting the entire system is an ongoing validation and quality assurance framework.

4 The Framework that Surrounds the PCR Reaction Affects Performance

The system that surrounds the polymerase chain reaction (PCR), including the input steps leading to the PCR, the processes for addressing the results obtained and their interpretation, and a comprehensive framework of quality assurance (QA) are critical to overall system integration and to the confidence users have in a PCR assay in an operational context.

SAMPLE COLLECTION, EXTRACTION, AND PURIFICATION AFFECT PCR ASSAY PERFORMANCE

The results obtained through PCR amplification and detection form the basis for subsequent decision making by the users who participate in BioWatch. Characterizing the assays to ensure that they meet performance standards is important to the success of the program. However, even the most perfect assay will be limited by the material that serves as its input. Aerosol samples contain a mix of viable and nonviable microorganisms along with numerous other substances (dust, pollen, dander, etc.). Sample collection, extraction, and purification steps precede the PCR and will affect the sensitivity and specificity of the results that can be obtained (Figure 4-1). This section discusses some of the issues that these steps present for PCR analysis.

Aerosol samples generally have a lower biomass for a given volume than other types of environmental samples, such as water or soil samples. The types of sample collection system used by a biomonitoring

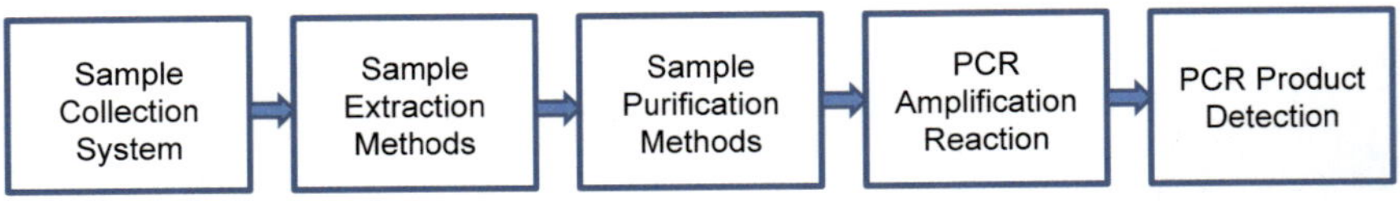

FIGURE 4-1 PCR Assay. Several key steps precede the PCR amplification reaction. When combined with a detection mode, these steps form a complete PCR assay.

program affect the air volumes that can be sampled within a defined period of time and the types of downstream analyses that can be conducted. In the current BioWatch system, air flows across a dry filter, and particulate material above a certain size threshold will be deposited. This system results in desiccation of organisms that become embedded in the filter. Although biological agents that form spores are environmentally hardy and may remain viable, this type of airflow system is less likely to allow recovered microorganisms to be cultured to determine whether they were live at the time of their collection. On the other hand, BioWatch aerosol collector technology uses a relatively large flow rate, enhancing the ability to sample a large volume of air. Other types of potential sample collection technology, such as impaction onto agar plates or into liquid collectors, also exist and would have their own associated advantages and disadvantages for the program. Once a BioWatch filter has been collected, the sample preparation steps (extraction and purification) critically affect the subsequent PCR.

Introduction of Sample Biases

Common extraction methods include soaking samples to release embedded material, often accompanied by mechanical agitation or homogenization (e.g., bead beating) to break up components such as cell walls, and freeze/thaw cycles and detergents to lyse cell membranes and release nucleic acid. Different types of microorganisms will be extracted more or less easily from filters, and may be more or less easily damaged during the extraction process. Spores are hardy, for example, but may require aggressive techniques to break them open and release sufficient amounts of an agent's DNA. Gram-negative bacteria are more easily lysed, but their genomic material also may be more easily sheared and degraded during extraction. As a result, the specific extraction methods used have the potential to bias the types of organisms that will be most readily available to the PCR amplification and thus be most efficiently detected.

Ability to Remove Inhibitors

A wide variety of substances may be co-extracted with sample nucleic acids. These substances may act in a number of ways to inhibit the effectiveness of a PCR assay. For example, an inhibitor may interfere with cell lysis and reduce the release of target nucleic acid. Once the nucleic acid is released, an inhibitor might degrade DNA so that it is not useable for PCR or might form a complex with it so that the primers, probes, and polymerase cannot access it. Alternatively, an inhibitor might block or affect the functioning of the polymerase enzyme during the PCR, reducing or preventing DNA amplification. For fluorescence-based real-time PCR, contaminants in the reaction also can interfere with successful fluorescence detection.

Substances that have been reported to function as PCR inhibitors include heavy metals and a number of organic or inorganic chemicals. The complex environmental backgrounds expected to be present in BioWatch filter samples increase the potential that inhibitors may be present. For example, the presence of tannic acid, which can be found in some types of leaf litter, and humic acid, which can be found in soils, has been reported to inhibit PCR reactions (Opel et al. 2010; Schrader et al. 2012). The presence of inhibitory substances in environmental samples has been reported to decrease assay sensitivity by an order of magnitude (Wilson 1997). On the other hand, Peccia and Hernandez (2006, p. 3951) report that "[r]eview of PCR application in aerosols suggests that, detection levels of $10–10^2$ gene copies can be achieved provided that appropriate steps to mitigate inhibition have been taken and the cell lysis and DNA purification steps in sample processing are efficient." Taking into account extraction inefficiencies, detection limits of 2,000-3,000 Gram-positive bacterial cells and 10-25 fungal spores per aerosol collection filter have been reported in the literature (Hospodsky et al. 2010).

As a result, the performance of PCR assays suitable for use in BioWatch needs to be tested against panels of environmental substances and in the context of likely environmental background (such as previously tested filters). The potential impact of inhibitors also informs some of the necessary controls during the PCR, such as inclusion of a control to demonstrate that the reaction mix can successful amplify a ubiquitous DNA target such as the 16S ribosomal RNA gene of bacteria. Various purification steps can be used to separate extracted nucleic acid from other substances prior to use in PCR. For example, extracted DNA can be treated with phenol and chloroform to separate it from co-extracted proteins. However, there will be sample losses and inefficiencies during all of the sample preparation steps.

The importance of sample extraction and purification steps to the sensitivity and effectiveness of downstream PCR raises two issues for consideration by BioWatch:

- Protocols for sample preparation must clearly define the steps to be conducted, reagents and methods to be used, and associated parameters such as times and temperatures.
- BioWatch collectors are deployed in different locations in different jurisdictions across the country, and some are located in indoor settings while others are sited outdoors. It may not be possible to define a single procedure for sample extraction and purification that works for all of these collectors under all circumstances. Rather, a suite of sample preparation protocols may need to be defined along with an accompanying decision framework to inform the determination of which procedures to use where and when.

The BioWatch program provides sample processing and extraction protocols to the jurisdictional laboratories to be used on collected filters. Two extraction options are provided—a protocol using a single tube filter apparatus and a protocol using a 96-well filter apparatus for multiple samples—along with a recommendation for which to use based on the number of samples the laboratory processes. An additional protocol has more recently been provided for high pollen counts; its use is at the discretion of the jurisdictions. The program is considering opportunities to improve these protocols, and it has indicated that such efforts are separate from the PCR assay test and validation that are the subject of the present study. The critical nature of sample preparation steps for PCR assay results is why guidance documents such as those from Food and Drug Administration (FDA) or the Minimum Information for Publication of Quantitative Real-Time PCR Experiments (MIQE) guidelines include sample preparation as part of the assay submission information. This report does not delve in further detail into these PCR input steps, because they were not the focus of the statement of task. However, their importance should be strongly emphasized and be part of any validation of a PCR assay.

Optimization of Reaction Conditions

The reaction conditions during PCR affect performance and may need to be adjusted during assay development. The enzymes used in PCR will polymerize available nucleic acids either perfectly or imperfectly from whatever material is present in the reaction mixture. It may seem counterintuitive, but a PCR can have too much DNA or other nucleic acid in the sample. As a result, copies of nucleic acids from an imperfectly matching template can be synthesized by the polymerase, and could be detected as a false positive in some detection formats. This is particularly true when there is an excess of template in the reaction. In other circumstances, too much DNA in the reaction can result in little or no amplification. Reagent balancing is key to PCR and any synthetic reaction. The unknown environmental background in input samples contributes to this issue because background content cannot be controlled. Nucleic acid will be extracted from multiple unknown organisms that make up the material collected on a BioWatch filter. Measures of pre-PCR DNA concentration in a sample should be taken prior to use in a PCR reaction. Simple tests such as use of fluorescent intercalating dyes can be used as a crude DNA sample content quantification method.

Detection Technology

The technology used to detect the amplified PCR products can also affect assay performance. Such detection parameters include the method of detection (e.g., SYBR green DNA binding dye, Taqman fluorescent probes, and hybridization to beads) and the instrument platform (e.g., Applied Biosystems, Roche, or others). In the case of hybridization of PCR products to beads, which is one strategy used in multiplex PCR assays, the PCR products and DNA conjugated to the beads must have high complementarity, resulting in some potential restrictions on the primers and target sequences used. This is an example of a situation in which the detection method can affect PCR requirements. To understand PCR assay performance, the performance must be evaluated on the detection platform that will be used. When an assay has previously been validated on a particular platform and a change is contemplated, bridging studies must be conducted to re-test the assay on the new platform.

VALIDATION IS THE FRAMEWORK THAT SUPPORTS QUALITY

Just as steps that provide input to the PCR are critically important to the overall PCR assay performance, so too are the interpretations that follow a PCR result. As shown in Figure 4-2, a result can lead to the determination that a pathogen's nucleic acid is present on a BioWatch filter, resulting in further downstream actions, or that a pathogen's nucleic acid is not present, in which case routine operations continue. Surrounding the entire system from sample collection through post-result decision making, however, is a QA and validation framework. This framework supports robust feedback to the PCR assay (and other steps in the process) in order

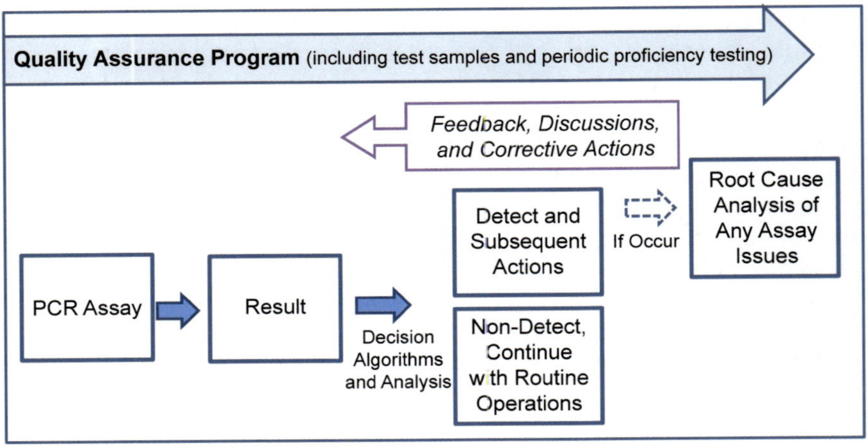

FIGURE 4-2 PCR assay performance is influenced by feedback from the steps that follow a result and the surrounding QA infrastructure.

to further inform the understanding of assay performance, detect issues that arise, determine the root cause of such issues, and implement appropriate corrective action plans.

The Process of Validation: General Principles

Validation is the process of assuring and documenting that a thing, such as a test, device, or process, fulfills the purpose for which it is intended. As defined by Budowle et al. (2014, p. 4), validation accomplishes the following goals:

1. Assesses the ability of procedures to obtain reliable results under defined conditions;
2. Rigorously defines the conditions that are required to obtain the results;
3. Determines the limitations of the procedures;
4. Identifies aspects of the analysis that must be monitored and controlled; and
5. Forms the basis for the development of interpretation guidelines to convey the significance of the findings.

A full validation pipeline includes multiple steps, from defining intended purpose through post-implementation assessment. Validation is not performed one time, prior to implementation, but is a continuous process to ensure that the test is performing as expected (i.e., part of the continuous quality control/quality assurance [QC/QA] process). Using validated approaches is a critical part of providing confidence in the results to the program's users. In the context of BioWatch, validation provides an understanding of the strengths and limitations of the data obtained from the PCR assays, which is crucial in interpreting results and using them as a basis for taking action. It is important to note that validation is not the same as setting the performance standard for an assay. The performance standard sets the criteria that the assay must fulfill in order to be able to meet its intended purpose. Validation is the process to ensure that a given assay meets or exceeds those requirements. Extensive guidance exists on the basic validation process and criteria for validating diagnostic assays designed to detect microbial agents for human disease (CLSI 2004, 2012; Jennings et al. 2009; Burd 2010), veterinary infectious diseases (OIE 2013), pathogen detection in food (FDA 2011), and law enforcement investigations using microbial forensics (Budowle et al. 2008, 2014).

Validation includes, but is not limited to, characterization of the analytical performance of the assay when used according to a given preparative method and instrument. As discussed in Chapter 3, measures to characterize the performance of PCR assays include, but are not limited to, limit of detection, sensitivity, specificity, repeatability, and robustness and are usually carried out by one or more independent laboratories to complement the characterization work done by the assay developer. In addition to analytical validation, validation also undertakes characterization of the performance of the assay in an operational context. For veterinary or clinical diagnostics use of a real-time PCR assay, factors such as the assay's diagnostic sensitivity, diagnostic specificity, and positive and negative predictive value will be examined (OIE 2013).

The series of steps in the validation framework, for real-time PCR detection assays applicable for BioWatch, are captured in Figure 4-3. As can be seen in the figure, after studies conducted by the developer and independent laboratories to validate assay performance characteristics, verify that user labs can obtain appropriate assay performance, and validate and verify the assay in a system operational context, validation includes ongoing monitoring of assay and operational performance. This monitoring includes laboratory proficiency testing, which helps to ensure that samples are handled and assays performed correctly according to procedures. In addition to obtaining sample results, this testing may deal with aspects such as sample entry, report writing and interpretation, and technical and administrative review. But a robust QA program does more than this. The data obtained feed back to inform the understanding of assay performance over time. Assessing the performance of positive controls over time can be an invaluable way to monitor the assay to pro-

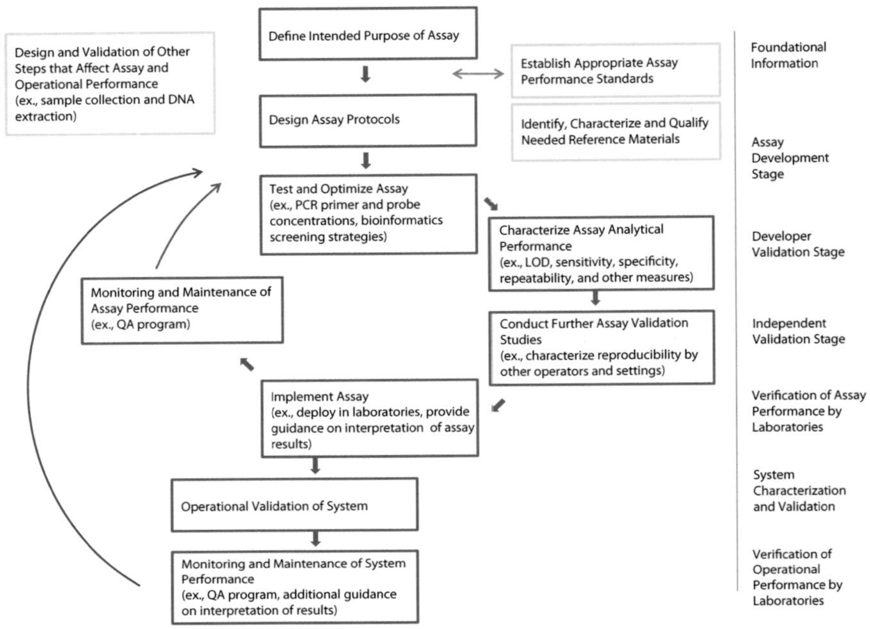

FIGURE 4-3 Steps in assay and system validation.

vide such feedback. The results may influence changes to assay design and to the strategies used for testing and validation. The data also serve to inform an understanding of assay and system performance limits and guidance around the interpretation of results for decision making.

CONTINUOUS MONITORING AND ASSESSMENT IS A NECESSARY PART OF VALIDATION

The Role of Quality Assurance

QA is a comprehensive infrastructure that defines the overall laboratory system including validation, maintenance of standard operating protocols, training and testing standards, proficiency testing using quality-controlled reference materials, audits, and corrective action plans.[1]

[1] Quality assurance and quality control are related but distinct concepts. Quality assurance is a comprehensive system and infrastructure that ensures that the laboratory meets standards of quality. Quality control, on the other hand, focuses on the performance of specific assay testing activities.

In 2011, the Department of Homeland Security (DHS) implemented a QA program for the BioWatch program. The program is based on ISO 17025 standards and includes an analytical component and a field component, with yearly reviews. The QA program includes (1) plans and audits, which define requirements and monitor compliance with those requirements; and (2) performance assessment, which evaluates whether performance is at a level to meet program objectives.

The performance assessment incorporates data from the BioWatch jurisdictional laboratories arising from daily, external QC samples that are sent to the laboratories as well as the internal QC samples that are routinely run with the assays. Each laboratory is required to track its data using control charts. An external contractor, Signature Science LLC, evaluates the data acquired from the laboratories to determine if the laboratory is meeting the established metrics and goals; data are examined by Signature Science and DHS based on 8-week sliding windows and each laboratory is reportedly audited every 2 years with the findings provided to the laboratory and to DHS. Information associated with the BioWatch program, including data obtained through the QA program, is available to laboratory directors through a BioWatch portal. In general, a BioWatch laboratory has access to its own data and to trend data compiled across the network, but not to specific data obtained from other laboratories. Proficiency testing is also conducted 3 times per year. As with the BioWatch screening assays, the Department of Defense (DOD) Critical Reagents Program (CRP) is the source for the reagents, QC samples, and proficiency test samples. Benchmarks have been established for false positive and false negative rates (starting from the filter extraction step), which, when linked to confidence intervals for those error rates, determine whether performance goals are being met. This evaluation strategy provides performance reviews of gene targets across the various testing laboratories and of various laboratories across gene targets (Isbell, 2014; Walter 2014b).

Assay Ruggedness

Data from the QA program provide information that can inform user understanding of BioWatch assay performance. For example, the robustness or ruggedness of an assay refers to the ability of the method to achieve acceptable performance under the variable conditions that come with routine use (e.g., the use of different batches of reagents, performance of the assay by different operators in different laboratories, or small instrument-to-instrument differences even when users are running the assay on the same instrument platform). Data obtained through the QA program enable monitoring for changes or variations in the false positive and false negative rates obtained during QA and proficiency testing, which can provide a multiyear assessment of how robust the assay is.

The continuous performance monitoring stage is also the time during which real-world assay performance issues are likely to arise that may have significant effects on the confidence that jurisdictional decision makers have in BioWatch results. The system can help identify developing issues in the performance of an assay that may need to be addressed through changes to the assay signatures, primers, probes, strain panels, or other features (requiring new performance testing) or through changes to the decision-making algorithm.

Signature Maintenance

One of the practical implications of the increasing genetic knowledge and recognition of microbial genetic diversity, discussed in Chapter 3, is the need for maintenance of the signatures used in the assay and the existence of signature erosion. As the amount of genetic information continues to increase from the identification and sequencing of new strains of a species and of near-neighbor organisms, signatures that were once thought to be uniquely identifying for a particular strain may no longer be, in fact, unique. Such results lead to erosion of the utility of the signatures used in PCR assays, but also provide an opportunity for refinement and improvement of the system to be more selective for the specific strains that pose a threat to public health. As a result, a real-time PCR assay for pathogen detection is not a static, one-time process of development, validation, and deployment.

At regular intervals, the signatures used in such assays must be screened *in silico* against databases of genomic sequences to flag potential issues such as cross-reactivity with newly identified near neighbors or for the inability to detect a new pathogenic strain added to inclusivity panels. This testing can identify potential false positive and false negative situations, with follow-up laboratory testing carried out if needed. Lack of proper PCR signature maintenance can lead to overconfidence in signatures because "knowable" false positive and/or false negative events have remained unknown. The need for periodic *in silico* and laboratory re-testing of expected assay performance should be part of an assay performance standard suitable for BioWatch. *In silico* testing should be undertaken at least annually.

Other implications for assay performance standards are

- Each tested BioWatch agent is likely to require a different number of strains of the agent and of near neighbors to enable adequate testing. One size definitely does not fit all when it comes to testing these PCR assays.

- Both the sequence of the signatures used in the assay and the composition of appropriate inclusivity and exclusivity strain panels can be affected as new microbial genetic knowledge is obtained.
- Appropriate strain panels for highly diverse organisms such as *Burkholderia* spp. may be significantly different when the assay is used for screening versus for secondary testing. For example, a Stakeholder Panel on Agent Detection Assays (SPADA) working group previously developed a strain panel with 60 well-characterized strains (AOAC 2011 in preparation-a,b). However, the chair of that panel provided the caveat that a final PCR assay used for confirmation would need to be tested against over 1,000 strains to ensure its ability to correctly distinguish strains (Keim 2014).
- A clear process is needed for conducting the regular assay re-testing required and for updating versions of the assay and the performance standard so that signature erosion is clearly documented.

The issue of microbial genetic diversity and how to account for it is relevant for any nucleic acid–based detection technology, including the real-time PCR assays currently used by BioWatch. As more information is obtained about the numbers and types of microorganisms in the environment and as increasing amounts of sequence data for these microbes become available, assay design, performance characterization, and validation become more complex. The implications of microbial diversity and the potential for fewer good signatures and greater numbers of panel strains also provide an incentive for programs such as BioWatch to look at the applicability of advancing technologies such as next-generation sequencing (NGS), discussed in Chapter 5.

APPLYING VALIDATION PRINCIPLES TO ASSAY PERFORMANCE CHARACTERIZATION: PUTTING THESE INTO PRACTICE

Chapters 3 and 4 discuss the validation process and principle steps in establishing PCR assay performance. The following section illustrates how these could be applied to BioWatch to provide a reasonable amount of statistical confidence in the results obtained through laboratory assay characterization and validation, while seeking to keep numbers of replicates and numbers of strains practical. The approach discussed below does not duplicate the full level of detail contained in guidance documents such as the AOAC validation guidelines, SPADA Standard Method

Performance Requirements, Public Health Actionable Assay (PHAA) and Federal Standards for Assay Performance and Equivalency (FSAPE) documents, and other sources, which necessarily contain comprehensive and specific explanations for the ways that tests are carried out and data is handled and reported. These represent valuable resources to consult in fleshing out the details of a BioWatch standard.

The following process represents an approach that can serve as a starting point. The basic dimensions are given below and are explored in an example. It is important to emphasize that the specific details of the process ultimately used by the BioWatch program will need to be determined by the program and established in consultation with its stakeholder community.

The committee also noted that the list of pathogens for which BioWatch tests is based on risk analyses and so far has not included RNA-based agents. Should RNA agents be included in the future, the standards for assay validation would be similar to those for agent detection assays for DNA-based microorganisms. In such a situation, the reverse transcriptase (RT) enzymatic step would be considered part of the PCR reaction and the need to optimize and test reaction conditions for potential inhibition also would apply to the RT enzyme.

Starting point for standards approach recommended by the committee:

- *Determination of limit of detection.* A reasonable approach to determining an assay's analytical limit of detection (LOD) in a laboratory setting is to conduct serial dilution at a range of concentrations bracketing the estimated LOD. This reflects guidance on LOD determination provided by the Clinical and Laboratory Standards Institute (CLSI), and is also in line with assay data packages through DOD and the PHAA and FSAPE documents.
- *Expanded* in silico *testing.* Use of *in silico* screening presents a relatively fast and low-cost way to predict assay performance using as many strains as are available as high-quality sequences in reference databases. It is also possible that *in silico* screening may uncover reference database curation issues. These could include pathogen DNA erroneously incorporated into host genomes (or vice versa), contaminated or mixed pathogen genomes, or misidentified pathogen genomes. The lack of ground truth must always be considered when evaluating DNA signatures for pathogens.
- *Strain panels to use for testing.* The panels recommended by SPADA represent a good starting point and were developed with stakeholder participation. All strain panels need to be regularly reviewed and updated to account for new genetic knowledge. As

a result, it should be recognized that the current SPADA panels may not provide sufficient genetic coverage for certain pathogens without additional review and modification.

- *Testing of inclusivity and exclusivity strains*. The strategy used by Los Alamos National Laboratory (LANL) for testing detectability or nondetectability of inclusivity and exclusivity strains represents a reasonable model for BioWatch.
- *Cost.* Initial costs for assay development and primary performance characterization should be undertaken by the developer. Independent validation should subsequently be conducted by DHS.

This approach is spelled out in greater detail in the example, below.

Hypothetical Example: The BioWatch program decides to deploy a new set of assays against a pathogen of interest.

Assay Development (for a New Assay): The assay developer selects the regions of the tested pathogen to be targeted for detection and designs associated primers, probes, PCR conditions, and other aspects of the assay protocol.

- The developer uses best practices in PCR assay design and optimization to create one or more assays that will detect the pathogen of interest under appropriate reaction conditions, while not cross-reacting with organisms that are not of interest. As part of this process, the developer undertakes *in silico* assay performance testing by screening for predicted inclusivity, exclusivity, and environmental background performance against organisms whose high-quality sequences are available in reference databases. This approach presents an opportunity to anticipate performance against as many relevant strains or organisms as possible.
- If the sequences are available, *in silico* performance predictions should be examined against the full set of strains and organisms recommended in the PHAA panels (under the section describing testing of environmental samples, not clinical samples).

Initial Performance Validation: The developer, or a laboratory contracted by the developer, determines assay parameters such as limit and dynamic range of detection, sensitivity, and specificity, along with associated confidence limits on the results.

- The developer obtains the necessary nucleic acids for the strains that will be tested from a repository that is able to supply quality-controlled materials. For BioWatch, this repository could be the CRP. If necessary materials could be provided by sources or repositories other than CRP. DHS will need to ensure that those conducting validation testing have sufficient access.

- A primary reference strain and three to five additional reference inclusivity strains are used in determining several of the performance characteristics. These strains should be specified in the performance standard and selected from among the inclusivity strains that the assay should detect.
- The developer determines the LOD for the primary reference strain. If there is agreement on a specific acceptable minimum detection level to be fit for purpose (AMDL, in the case of a standard such as SPADA), testing to determine probability of detection around that concentration could alternatively be undertaken.
- A reasonable approach to determining an assay's analytical LOD in a laboratory setting is to conduct serial dilution at a range of concentrations bracketing the estimated LOD, using $n = 60$ replicates with acceptance criteria of at least 58/60 for a given concentration, followed by appropriate curve fitting. If the assay is able to detect 60/60 samples, there is an approximately 95 percent probability that the assay will detect that quantity of DNA with an associated lower confidence limit of 95 percent. If the assay is able to detect 58/60 samples, there will be an approximately 90 percent long-term probability of detection at that concentration of DNA, with an associated lower confidence limit of 95 percent. This approach represents a fairly stringent characterization of assay performance for this variable—as noted earlier, for example, the long-term probability of detection at a particular quantity of DNA must be 99 percent in order to achieve 60/60 detections with greater than 50 percent probability (DHA 2013a).
- A smaller number of replicates could first be used to narrow the expected range within which the LOD falls, with a larger number of replicates used around the hypothesized LOD (an approach that is used in PHAA and FSAPE guidance).
- The developer tests the LOD of the three to five additional reference inclusivity strains at a subset of values around the calculated reference strain LOD to examine detection variability across strains.
- Targeted laboratory testing is undertaken to validate the predicted inclusivity, exclusivity, and environmental background performance against a prioritized subset of strains based on morbidity or mortality, transmissibility, and host range (inclusivity panel) and on sampling relevant genetic and environmental background diversity (inclusivity, exclusivity, and environmental panels).

The SPADA strain panels represent a reasonable starting point, updated as needed to reflect current genetic information and prioritization since the panels were developed.

- Characterizing detectability of inclusivity strains or nondetectability of exclusivity strains and environmental organisms by testing $n = 20$ replicates, followed by testing 20 additional replicates if an unexpected detect or nondetect result is obtained, with acceptance criteria of 20/20 or 39/40, represents a reasonable model. If the assay is able to detect 20/20 samples, the 95 percent lower confidence limit on the probability of detection of the reference strain at this concentration will be approximately 86 percent. If the assay is able to detect 39/40, the 95 percent lower confidence limit on the probability of detection of the reference strain at this concentration will be approximately 89 percent.
- In addition to LOD and specificity tests conducted using purified nucleic acid, tests performed in the presence of realistic background matrix, such as by spiking DNA onto used BioWatch filters, is conducted to investigate whether assay performance is affected. For example, a reduced signal could indicate inhibition due to the presence of substances such as metal ions or humic acids. An understanding of potential inhibition due to background is important is assessing negative results. Initial assay robustness also is conducted by the developer to understand performance in the context of variable conditions that may be encountered in use and that may impact PCR. Appropriate positive and negative controls should be included in all of the testing, as discussed in Chapter 3.
- The developer should report the data obtained and accompanying calculations in an acceptable format so that it can be understood and used by those who will perform the tests, those who will interpret the results, and other stakeholders that may have to respond based on the results.

Independent Validation: DHS undertakes or contracts with an independent laboratory (not the developer) to validate the performance information on the assay. Similar steps are repeated, but the focus is on targeted testing of the documented assay performance.

- The LOD is validated using nucleic acid with an appropriate number of replicates (see above). In this case, testing should be conducted in the context of the expected sample matrix.
- Validation testing is conducted to confirm the assay performance reported by the developer against the strains used in the initial

laboratory inclusivity, exclusivity, and environmental background testing (see above).

- BioWatch uses certain assays for screening and other assays for secondary verification that a tested pathogen's nucleic acid is present on a collected filter. Although the same core performance standard and validation framework is appropriate for these PCR assays, "tiers" of validation testing could be implemented that emphasize different parameters based on intended use. Laboratory testing using the full sets of PHAA panels is not required for a screening assay, although current SPADA panels may not capture sufficient genetic diversity for all pathogens, and all panels must be revisited regularly. Laboratory validation for a secondary assay could include testing of additional inclusivity, exclusivity, and environmental strains (e.g., drawn from PHAA/FSAPE panels) in order to sufficiently characterize assay sensitivity for this type of intended use.
- Alternatively, a set of subsequent assays (for use following the screening assay) that draw on PCR, NGS, and other technologies could be established for each tested BioWatch agent and designed to serve different roles in ruling in, ruling out, confirming, or characterizing the sample.

When Pulling an Existing Assay Off the Shelf for Deployment: A relevant assay and accompanying package of performance data may have previously been developed, perhaps for another purpose, but not deployed by BioWatch. The validation data generated for that assay could be used to support the use of the assay. In this case, new genomic information may have been obtained since the assay was designed. An *in silico* reexamination of expected assay performance is carried out to look at the target region of amplification, primer and probe sequences, and anticipated performance against inclusivity, exclusivity, and environmental strains in order to check for potential performance issues prior to conducting additional validation testing as required using the framework described above.

System Characterization and Validation: Any assay must work in concert with all of the steps in the system from sample collection to data analysis and ongoing quality assurance. Similar to the independent validation stage, above, targeted testing to characterize performance should be conducted in a more realistic operational context (e.g., to determine LOD and performance against inclusivity, exclusivity, and environmental background in an operational system context).

- Testing is conducted using samples in the context of the collection media, matrix, and realistic environmental background.
- Sample extraction and preparation procedures are tested from realistic environmental background to determine performance; changes to sample extraction, preparation, or assay procedures made as necessary.
- Assay performance is tested against a panel of environmental substances that might interfere with PCR amplification and detection. SPADA environmental substances panels represent a reasonable starting point for testing. Utilizing BioWatch filter samples from jurisdictions that have experienced PCR inhibition problems in the vicinity of particular BioWatch sampling units would also be useful.

Verification of Operational Performance by User Laboratories: The performance of the assay (either in isolation or, more appropriately, in the context of the system from sample collection to output) should be tested by the BioWatch jurisdictional laboratories that will use the assay. This testing indicates whether laboratories across the program network can achieve the anticipated assay performance results and gives a sense of the operational robustness of the assay.

- This stage determines whether there are local issues (e.g., PCR inhibition, environmental cross-reaction with some or all assays for the organism due to unknown near neighbors, etc.) that may arise and, if so, they will need to be addressed.
- Verification of LOD could be conducted using $n = 20$ replicates; verification of detectability or nondetectability of the prioritized inclusivity, exclusivity, and environmental strains could be conducted using $n = 5$ replicates for each tested strain.

Monitoring and Maintenance: Validating and deploying an assay is not a one-time event.

- Operational performance by user laboratories should be monitored over time through a QA program, which may flag potential issues that develop in particular locations or under particular conditions or from lot-to-lot variation.
- New genetic knowledge of relevant microorganisms and of environmental background is rapidly being acquired. DHS should reexamine each assay at regular intervals, ideally annually. This review should include *in silico* testing of expected assay performance for the target region of amplification, primer and

probe sequences, and anticipated performance against inclusivity, exclusivity, and environmental strains as well as all other relevant microbial and host genomes available in databases. If potential issues are detected, targeted laboratory testing should be conducted to determine whether the assay continues to meet performance goals or needs to be redesigned. The results of the annual *in silico* and periodic laboratory re-testing should be documented, as well as the action(s) taken, in a reasonable time frame when the need for assay redesign is indicated.

CONSTRAINTS OF THE CURRENT SYSTEM AND OPERATIONAL EXPERIENCE OF BIOWATCH JURISDICTIONS

Building on the detailed discussion of analytical characteristics that define the performance of real-time PCR assays and the factors that need to be considered for a BioWatch-suitable standard, it is important to return to the real-world context in which these assays are used and the experiences of BioWatch jurisdictions. It is the laboratorians in the network of state and local public health laboratories who have to be confident that they sufficiently understand assay performance and have confidence in the results and the local and national BioWatch Advisory Committee members who need to use assay results in actionable decisions.

The implementation of the BioWatch QA program is a positive recent development to provide guidance. The BioWatch program also conducts periodic workshops or webinars for the network of jurisdictional laboratories in order to discuss the obtained performance data and appropriate interpretation. The QA program thus is a critically important component of the program management processes. However, the committee observed several areas during its data-gathering where it appears that user jurisdictions would benefit from additional discussions and information. An increased effort in communication, education, and training would be invaluable for interpretation of results and building user confidence. Information from these discussions also needs to be translated back to the public health officials who make decisions based on the laboratory results, in order to inform their understanding of the information they receive.

Additional Parameters Influencing Decision Making on Assay Results

To support actionable decision making based on assay results, it is helpful to examine an assay's sensitivity, to the extent possible, in the context of the pathogen dose required to cause disease. For example, if an assay is able to detect a single bacterium with sufficient confidence, but the infectious dose of the pathogen would require exposure to at least 1000 bacteria and the assay result yields a threshold cycle (C_t) value around the cutoff cycle (therefore, barely above the threshold for judging it a positive result), a jurisdiction may respond differently. Likewise, if the same assay has an LOD of 100 bacteria (a less sensitive assay), but is producing a very strong signal, the laboratory may respond rapidly and aggressively.

The interpretation of assay results in this context is significantly complicated by the fact that it is extremely difficult to link an assay's analytical sensitivity with the aerosol quantity of a pathogen. If a BioWatch collector was sampling the very edge of a plume, it may indicate only a weak signal when in reality the center of the plume could have clinically significant pathogen concentrations. So it is inherently problematic for jurisdictions to disregard weakly positive results if they are over the threshold for a positive detection. This is also an opportunity to raise the question of guidance on whether and how to draw on the semi-quantitative information provided by the real-time PCR curves and resulting C_t values, as opposed to yes/no decision making based on cutoff C_t thresholds. Because the screening assay tests for one signature and the secondary verification or confirmatory assay tests for several additional signatures, situations also can arise in which some signatures are positive and some are negative. The program has decision algorithms (e.g., that all 3/3 signatures must give positive results to declare a BAR), but the committee's impression is that jurisdictions would like to have further discussions on what these types of "partial" results truly mean. The committee agrees. Discussions among the BioWatch program, federal experts, and jurisdictional technical experts on additional guidance on assay interpretation for situations beyond simply detect/nondetect could help inform operational decision making.

Troubleshooting Assay Issues Encountered by Jurisdictions

The operation of the QA program and interpretation of the resulting data could be even more closely integrated with the network of BioWatch laboratories in order to better troubleshoot key operational issues that are encountered by individual laboratories and to provide guidance to the laboratories on the interpretation of results. For example, the issues encountered in jurisdiction "X" with relatively high rates of positive results in the assays for agent "C" do not involve false positive or false negative results on known QC test samples, but should be addressed as part of the QA framework because the issue involves overall performance and validation. The jurisdiction still needs a basis from which to address its specific situation. The reasons an assay does not work well in a particular jurisdiction could include

- Lack of understanding or knowledge of assay performance parameters, which are needed in order to understand assay limits and data interpretation;
- An environmental background issue in broad or limited regions within certain jurisdictions; and
- A technical issue with the performance of the assay.

The most effective solutions to address the different causes above will vary. Similarly, although a significant amount of useful data is obtained through the BioWatch QA program, state and local users may not always know what specific corrective actions have been implemented in response to issues identified in their own or in others' laboratories.

The impression that the committee obtained (recognizing that experiences may not be the same for all users and that there may be relevant information to which the committee did not have access) is that laboratories and jurisdictions address issues with assay results on an ad hoc basis. Incorporating a systematic mechanism into the QA system for feedback and for making and communicating decisions on whether to implement assay or operational changes would strengthen the already valuable effort. BioWatch needs to have a process to identify the root cause of an issue, discuss appropriate strategies to resolve it, and carry out processes to address it in an expeditious manner. For example, one suggestion might be modeled on the process of medical peer review, in which clinicians periodically come together to review cases and discuss what went right and what went wrong. The BioWatch program could consider holding sessions with federal and jurisdictional laboratory experts and public health decision makers to report on unusual assay results obtained during operational performance and discuss collectively how to solve them. If there is an environmental background detection issue in a jurisdiction, this might provide an opportunity to make a change to increase the specificity of the assay(s) in question that could eventually spread to other jurisdictions.

Communication Across the Program Network and with Additional Stakeholders

The QA operations and information largely involve the initial screening assay. The external contractor, who analyzes the data and provides the information to DHS for its further analysis and any decision making, does not have access to information from the Centers for Disease Control and Prevention (CDC), which is the source for the secondary assay. Because the initial and secondary assays currently are used as a two-step process by BioWatch laboratories, further communication and regular data sharing would be helpful.

BioWatch jurisdictional users also expressed the strong desire to have greater access to specific assay performance and validation data on both the initial screen and the secondary assay, and to have more detailed QA program data than are currently provided. The depth of laboratory technical expertise may vary from jurisdiction to jurisdiction, but in fundamental ways the BioWatch program operates locally. There are opportunities to better leverage the expertise that exists in, for example, the state and local public health laboratories. Laboratory directors and senior personnel in jurisdictions must present assay results and their interpretation to local BioWatch Advisory Committees and decision makers; these personnel express frustration in not having access to as much performance and validation data as possible to support their responsibilities (personal communications from jurisdictional scientists and officials, September 3-4, 2014).

This issue is not entirely within DHS's ability to solve. It will involve discussions with the jurisdictions but also with agencies such as CDC. Both BioWatch and the CDC LRN currently maintain user web portals as information repositories, which might be expanded to share additional data.[2] The need to maintain appropriate security around assay and operational performance information should be taken into account. Nevertheless, mechanisms must be implemented to convey relevant information while maintaining necessary security measures. Sufficient numbers of personnel in BioWatch laboratories should have (or have the ability to obtain) the necessary security clearance, and systems could be devised to share and discuss details of assay analytical and operational performance and validation. Greater transparency and data sharing would improve user confidence. Overall, the committee concluded that a good framework for the program is in place, but the importance of communication to confidence and actionability should be strongly emphasized.

[2] For example, the LRN portal includes data on the performance of LRN assays (such as LOD, sensitivity, specificity, and matrix) that resulted from testing conducted around 2008 by CDC using inclusivity and exclusivity strain panels assembled by CDC subject experts (this testing predates the current effort to submit LRN assay data to FDA) and would be available to LRN laboratory directors (Toby Merlin and Harvey Holmes, CDC, personal communication, December 2, 2014). The BioWatch portal similarly contains various data accessible to the directors of the state and local laboratories that perform BioWatch assays.

5 Looking to the Future: Multiplex PCR and Next Generation Sequencing

The changing technology landscape will affect the BioWatch program. This chapter comments on additional assay performance considerations that arise from the use of multiplexed real-time polymerase chain reaction (PCR) assays and potential implications for the program of continued advances in genetic sequencing technology.

MULTIPLEXING PCR ASSAYS FOR USE IN BIOWATCH

It is possible to "multiplex" a nucleic acid–based detection assay in various ways. Several examples of possible approaches include

- Parallel readout of multiple, singleplex reactions (not discussed further);
- Multiplex PCR using multiple, specific primer sets in a single reaction (see Box 5-1);
- Multiplex PCR using conserved primers in a single reaction to amplify a large number of related species for subsequent analysis (e.g., using primers directed toward 16S ribosomal DNA, which is capable of amplifying a wide range of bacteria, followed by sequencing of the amplification products to identify the organisms);
- PCR amplification using random, universal or whole genome amplification followed by another, high-throughput method of analysis and readout, such as via microarrays; and
- PCR amplicon, targeted probe enrichment, or metagenomic sequencing of various types.

BOX 5-1
Multiplex PCR

Multiplex PCR, as discussed in the report, is similar to the singleplex real-time PCR discussed in Box 1-1, except that each sample is designed to amplify and detect multiple target sequences rather than only a single target per reaction. Several detection methodologies exist. Target-specific fluorescent probes can be used that each emit at different wavelengths. Alternatively, probes specific to the different target sequences can be hybridized to beads having different florescence profiles (expanding the numbers of colors that can be detected). After PCR amplification, the DNA containing a mix of the sequences can be hybridized to the beads in order to detect the specific signal associated with each amplified target.

A

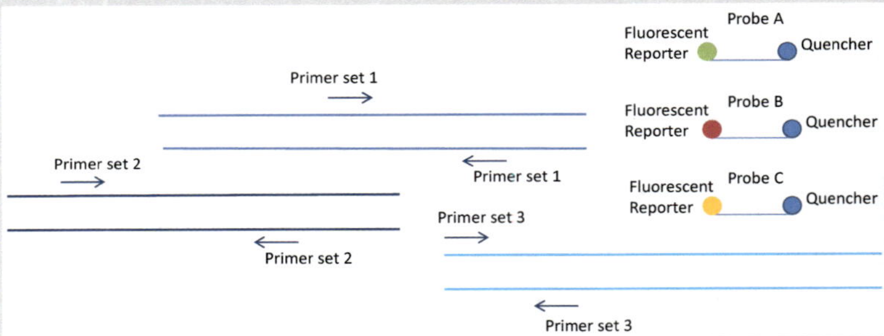

B

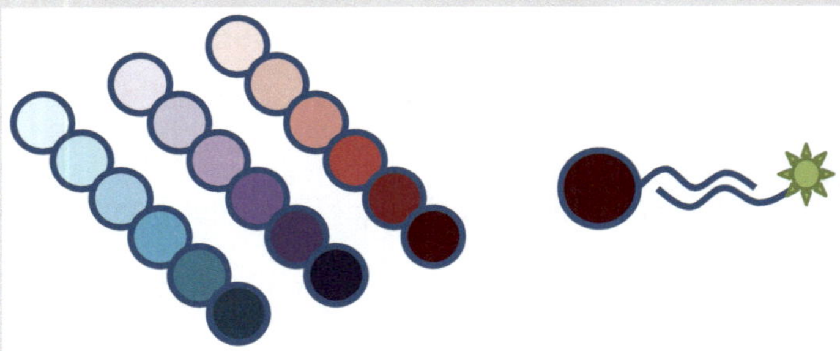

BOX FIGURE: Depiction of different strategies for multiplex PCR. **A.** Using target-specific fluorescent probes; **B.** Using fluorescent beads.

With several potential technologies, the signal obtained can suffer from nonspecific hybridization (e.g., microarrays), nonspecific protein signals (e.g., protein mass spectrometry), or limited resolution provided by information on only the relative percentages of A,C,G,T DNA nucleotides (e.g., nucleotide mass spectrometry). PCR with conserved primers is most useful for cases where it is desirable to detect a wide range of targets with broad sensitivity (e.g., the use of primers to detect 16S or *rpoB* sequences in bacteria [Case et al. 2007]). This method may be less useful for viruses, which are highly diverse, or specific organism identification to the species or strain level required by BioWatch for most targets. Sequencing can be prone to multiple potential sources of errors, some systematic (e.g., runs of homopolymers) and some random, and each sequencing technology has its own error model. Recent reports have examined potential alternative technologies for BioWatch detection (IOM and NRC 2014), and these are not explored in detail here.

The type of multiplex PCR technology that the committee focuses on in the report is the inclusion of multiple sets of target-specific primers in a single reaction (the second item in the list above and described in Box 5-1). All of the real-time PCR assays currently used in BioWatch are performed as singleplex assays, in which a single target DNA signature is amplified for each assay. The secondary assay for any of the tested BioWatch agents, which looks for the presence of three or more signatures, also is run in a singleplex manner, meaning that the "secondary assay" consists of three or more individual singleplex assays. However, if BioWatch assays were combined in a multiplex fashion, multiple targets could be tested at one time, potentially saving time and labor but with a potential loss of detection sensitivity. Multiplexing BioWatch assays has potential advantages and disadvantages for the program, some of which are listed below.

Characteristics of Singleplex PCR:

- Has good sensitivity and specificity as a result of carefully designed primer and probe sets. However, assay performance is dependent on availability of nucleotide information for primer design.
- Narrowly focuses on only one region of the genome. However, it may not provide sufficient discrimination between a biothreat agent and nonpathogenic flora because only a fragment of the genome is detected.
- Relies on conservation of primer binding sites used for amplification. Sequence specificity is required for primer and probe hybridization to the target. However, viruses have high diversity, which must be considered in designing primers and probes that

can detect the virus or that may need to anneal to a primer binding region that can exhibit variability. In bacteria, the potential dropout of targets on plasmids or in certain genes (such as the antibiotic resistance gene *mecA* in *Staphylococcus aureus*) may need to be addressed. As a result, the use of more than one target in an initial screen or in a secondary assay is beneficial.
- For serial testing, uses up a lot of sample, but there is potentially less input per target assay.
- Typically provides limited information on strain genotype and on additional pathogen characteristics such as antibiotic resistance.

Characteristics of Multiplex PCR:

- Targeted PCR has the same types of limitations as singleplex PCR (conservation of primer binding sites, narrow focus on specific regions of genome).
- "Random" PCR methods amplify everything in the sample ideally without biased representation of the targets. As a result, these methods can detect a broad range of targets, although a method to further type the targets is required.
- Multiplex methods must deal with interactions between multiple sets of primers and probes that may have different annealing temperatures and different performance in the assay, and that may cross-react and generate primer dimers, for example. An "all-in-one" approach may not be useful in the detection of many agents with large variations in diversity, genome sizes, AT/GC ratios, or other parameters that affect assay performance.
- Selection and use of proper controls can be challenging where many targets are being detected simultaneously.
- Rather than designing one multiplex to cover all desired targets, it is also possible to design several multiplex assays that each covers a set of targets whose performance under PCR conditions can be better optimized; such multiplex assays can be used in combination.
- Detection of a multiplex PCR typically requires more sophisticated instrumentation than a singleplex, because multiple colors need to be detected when using fluorescent probes (see Box 5-1). The excitation and emission spectra start to overlap when more than six or so probes are used in a reaction. Distinguishing multiple distinct colors can be a challenge to sensitivity and specificity of the assay. Options include use of hybridization to beads, or splitting up the assays into a combinatorial approach into many parallel reaction tubes for real-time PCR.

For nucleic acid–based multiplex options, such as the multiplex PCR focused on by the committee, the core technology includes a PCR amplification step. Thus, the considerations that guide the development of PCR assay performance standards for singleplex real-time PCR analysis, the focus of the rest of the report, remain relevant to the development of appropriate standards for multiplex assays. Multiplex assays will need to be characterized and validated for the limit of detection (LOD) of each target in the multiplex, and assay specificities will need to be determined by testing against appropriate inclusivity, exclusivity, and environmental panels. However, several additional considerations arise in the development and validation of multiplex PCR assays.

Just as in singleplex assays but on a larger scale, the inclusion of multiple pairs of primers and multiple probes into a single reaction mix can result in primer–primer interactions that interfere with target amplification or lead to the production of nonspecific amplification products and reduce assay sensitivity. The composition of DNA sequences also affects optimal thermal cycling conditions, and optimizing the overall reaction conditions becomes more difficult with an increased number of DNA targets. For example, some microorganisms have GC-rich genomes, and GC base pairs in DNA generally require a higher melting temperature because of the greater number of hydrogen bonds compared with AT base pairs. In a multiplex assay, reaction conditions need to be suitable for targets that may vary in composition. The nature of the target DNA and reaction conditions also can result in one target sequence preferentially amplifying over others; strategies such as reducing (or increasing) concentration of certain primers may be needed to enable sufficient amplification of all desired sequences. Development of a multiplex assay becomes a fairly complex matrix of variables, including cross-reactivity, the potential for dye quenching and spectral overlap, and the incorporation of appropriate positive and negative control references, and conditions will likely need to be adjusted and checked as probes and primers are added to the multiplex. In the case of hybridization-based detection assays with primers and probes on beads, the capture sequences must also be verified and adjusted as needed. As a result, it has been suggested that it may be easier to start from scratch in the design of a multiplex assay rather than to adapt several existing singleplex targets, primers, and probes to be run together in a multiplex manner (Naraghi-Arani 2014). This approach may be hampered if there is only one known signature to detect and discriminate a target. Despite these challenges, assay developers have successfully designed PCR assays that multiplex large numbers of targets. The BioWatch program indicated to the committee a particular interest in multiplexing the routine initial screening assays to save time and labor. Because the screening assays identify one signature per pathogen, the required BioWatch multiplex assay for this purpose would include fewer than 10 signatures in the sample.

Multiplex assays are becoming increasingly common and assay developers and regulators have substantial experience with optimizing and evaluating them. One of the first Food and Drug Administration (FDA)-approved multiplex assays was the Luminex xTAG Respiratory Viral Panel, which tested for 12 viruses and was approved for marketing in 2008 (FDA 2008). The field has continued to expand, and FDA now has guidance available on performance characterization for highly multiplexed nucleic acid–based diagnostics designed to detect 20 or more targets per multiplex (FDA 2014a). As noted, many core aspects of performance characterization and validation are similar for singleplex and multiplex assay development, but additional guidance provided by the FDA for highly multiplex reactions includes

- Demonstrating through *in silico* analysis and validation data that negative interactions among assay components such as primers, probes, and amplified targets are not interfering with performance.
- Providing information not only on the individual LODs for each target in the assay, but also on the assay cutoff values for each target and how these values were determined and validated.
- Using representative panels of organisms targeted by the multiplex assay to demonstrate system repeatability and reproducibility, in order to reduce the overall size of the required studies. Similarly, one or more representative targets of the multiplex assay can be used to demonstrate that carryover and cross-contamination do not occur in the device between samples, rather than testing all targets.

Finally, the current FDA guidance indicates that a change to an approved multiplexed assay, such as the addition of a new target, requires the manufacturer to submit a 510(k) premarket notification submission because it represents a new intended use of the assay. However, the guidance acknowledges that only a subset of performance studies would be required because much of the assay's previously reviewed performance is expected to remain similar. New data to be submitted would focus on the additional target (characterization of LOD and inclusivity and exclusivity detection, for example). Tailored studies on a representative subset of the existing targets also would be conducted, such as a limited series bracketing the LOD to confirm that it is performing consistent with the original assay version (FDA 2014a). FDA is concerned with clinical use of devices and reportedly considers the impact of assay changes on the potential false positive and false negative rates as they would affect making appropriate patient treatment decisions; thus performance data required by the

FDA can depend on the tested pathogen. The overall goal, however, is to ask developers to provide information necessary to evaluate performance without putting an undue burden on them (Hobson 2014). Standards approaches such as the Stakeholder Panel on Agent Detection Assays, Public Health Actionable Assays, and Federal Standards for Assay Performance and Equivalency also recognize that assays may be developed in a multiplex fashion; the documents that describe validation under these approaches generally indicate that multiplex validation studies should be conducted so as to test the individual PCR targets together in a single sample. These approaches in concert with the FDA guidance documents should provide a suitable basis for validation testing of multiplexed PCR assays for BioWatch.

The BioWatch program utilized actionable multiplex assays from late 2007 before withdrawing them from use in mid-2009. The proposed acquisition of Gen-3 autonomous detection systems (canceled in 2013) also was envisioned as operating with multiplex assays. When multiplex PCR assays were field-tested by the BioWatch program, concerns were expressed over loss of assay sensitivity (U.S. House of Representatives 2012b,c). The evaluation of BioWatch Gen-3 vendor technology undertaken by Los Alamos National Laboratory (LANL) used the multiplex PCR assays developed by the vendors. In the experience of LANL scientists, PCR assays run in a multiplex fashion did show reduced sensitivity compared with singleplex assays, but the effect was reportedly closer to a factor of 2 than to an order of magnitude. For example, if the LOD with an appropriate probability of detection (e.g., 95 percent) of a target in a singleplex assay was 20 copies per reaction, the LOD of the target in multiplex might be closer to 40-50 copies per reaction, a difference that falls within the potential range of experimental error (Kristin Omberg, LANL, personal communication, October 24, 2014). Given the relatively low numbers of copies detectable by the PCR assays, the statistical error bars that surround copy number quantification at low levels, and the uncertainty in moving from analytical characterization of an assay using purified nucleic acid to the real world use in the system with DNA extracted from aerosol filter samples, it is unclear whether this degree of difference in assay performance would have a significant operational effect.

As with other aspects that the committee examined, the primary challenge to the use of multiplexed assays by BioWatch appears to be a communication issue more than a technical one. It appears likely that suitable multiplex assays can be designed for use in BioWatch, particularly to replace the routine screening assays. However, data on multiplex assay performance and validation, compared with the existing singleplex assays as references, need to be shared with the laboratory experts in each jurisdiction in order to give them a basis from which to use the tests. The committee's understanding is that most jurisdictions did not have access to such data during the limited prior deployment of multiplex assays by the program, which contributed to uncertainly around use of the assays. Without providing performance data to relevant jurisdictional experts, discussing differences such as possible, but likely limited, reductions in sensitivity, and explicitly considering how any performance reductions would be likely to translate to system operational performance, it is not surprising that resistance to use of the assays occurred.

TAKING ACCOUNT OF STEP CHANGES IN TECHNOLOGY

The BioWatch program currently operates under several assumptions and limitations. These include the use of a two-stage screening and secondary assay process based on real-time PCR of a nominal number of signature sequences, which places the emphasis on making these PCR assays as reliable and robust (as sensitive and as specific) as possible. These assumptions are not necessarily unreasonable and the current system is functioning. Real-time PCR as the detection methodology makes sense for BioWatch's mission, although the committee has suggested that in certain cases additional rule-out assays could be performed.

Technology in the life sciences continues to advance rapidly, driven largely by medical applications. As BioWatch looks to the future, one particular technology to watch is next-generation sequencing (NGS). The advantage of NGS technology is that it can provide more rapid, less expensive, and higher-throughput sequence data from a sample. NGS protocols in current use still largely depend on amplification of the sample materials prior to sequencing, although amplification-free methods such as nanopore-based single molecule sequencing are being developed (Quick et al. 2014). The implementation of NGS technology by metagenomics (random shotgun) or whole-genome sequencing has the potential to provide more precise information on genotype or strain and opens the possibility of detecting engineered or novel biothreat agents with high sensitivity and specificity. For example, genetic profiles associated with antibiotic resistance or with virulence may be identifiable. In this way, the technology would enable BioWatch to expand beyond its current portfolio of defined inclusivity strains for a small number of pathogen species. In clinical diagnostic use, these types of sequencing analyses to identify the root cause of hard-to-identify infections are starting to be successfully employed (Wilson et al. 2014).

However, it also is important to be realistic about the challenges associated with this technology and how it might or might not be appropriate for BioWatch in the near term. The technology will take time to mature. For BioWatch, in particular, samples are extracted from a complex environmental background on collected air filters. Accurate taxonomic and strain identification (requiring sufficient sequencing coverage and sufficient length of assembled sequence reads) and separation of sequences of interest from the enormous unknown background would be both critical and computationally challenging. In fact, a useful assembly from unbiased metagenomic sequencing of BioWatch filter samples may not be possible. The curated databases and informatics required to undertake the necessary sequence assembly or mapping at this scale are developing rapidly but can still be considered immature. Furthermore, the BioWatch program is designed to be an early warning system; thus information needs to be obtainable with a relatively rapid turnaround time. Because much of the technology and informatics required are still under development and are not routine, the required performance validation (involving both the assay and the informatics pipeline) also would be complex.

How Sequencing Can Be Useful Now

One key way in which sequencing currently can be incorporated into BioWatch is as a tertiary, follow-up analysis method for cases in which jurisdictions experience assay-positive results that do not appear to be true biothreat detections. In particular, sequencing should be undertaken where jurisdictions experience repeated apparent detections of certain pathogens, in an effort to truly understand what is causing the results. The committee understands that sequencing by Centers for Disease Control and Prevention (CDC) laboratories from collected filters has been attempted in at least some such cases, but that an answer was not always obtainable, or the details may not have always been conveyed effectively back to the jurisdictional laboratories and officials. Metagenomic sequencing could also be used now as part of efforts to better understand background genomic diversity on BioWatch filters from various collectors in various jurisdictions and at various time points. As discussed in Chapter 3, certain microbial strains may produce positive assay detections as a result of time, weather, or location-specific environmental sources and improving the ability to predict these circumstances would aid users in interpreting results. The information obtained through sufficiently deep metagenomic sequencing would also serve as an especially valuable resource for the *in silico* design of assay target sequences, primers, and probes, which currently rely on available databases.

How Sequencing Can Be Useful in the Future

The committee agrees with the findings of a 2013 BioWatch workshop that NGS is not yet sufficiently mature to be considered as a replacement technology for the current PCR-based BioWatch operations (NRC and IOM 2014). However, the committee believes that NGS could play valuable roles for BioWatch in the near and far term.

Sequencing as Confirmation of a PCR Screening Positive

The primary issue is whether it is feasible to utilize metagenomic ("random" shotgun) NGS on a portion of a BioWatch filter to provide confirmation of a PCR screening positive result and to provide additional characterization of the material present in the sample. In this way, NGS could essentially serve as a replacement for the current real-time PCR secondary assays. Multiple factors are involved in calculating the feasibility of this approach, including

1. The amount of DNA that can be extracted from a BioWatch filter quadrant, because different NGS technologies require different starting amounts;
2. Whether deposition on the filter is homogeneous (an issue for the current assays as well);
3. Whether random amplification of the DNA is required to obtain sufficient starting material, which may introduce biases because some DNA regions may be amplified more efficiently than other regions;
4. The likely unknown extent of the genomic background on the filter and how this background relates to the genomic material of the pathogen(s) indicated as present by PCR;
5. The sequencing depth of the NGS technology used, in particular how many random reads and what read length can be obtained from complex mixtures;
6. The amount of pathogen present that NGS will need to be able to detect (e.g., the LOD for the NGS);
7. The cost of the NGS run;
8. The sample-in to sequence-out time;
9. The time needed to analyze the sequence data, which ranges from hours to days depending on the bioinformatics algorithms used and additional factors; and
10. Technology transfer requirements, especially bioinformatics.

Items 3, 4, and 5 on the list define the tradeoff between pathogen LOD and the sequencing depth needed. Because of the enormous genomic background present on aerosol filters, standard desktop NGS on an instrument such as the Illumina MiSeq (which currently produces approximately 15-24 million reads of up to ~250 base pairs in length) may support a LOD that is too many orders of magnitude worse than PCR for it to be useful for confirmation and further characterization. Deeper sequencing may take too long a time or may be too costly. Thus, while NGS is not feasible for routine BioWatch characterization currently, it may be acceptable as confirmatory follow-up to near-BAR events as discussed above. In the future, continued technology and informatics developments may make NGS technology suitable as a replacement for the BioWatch secondary assay.

A Note About Targeted Versus Metagenomic NGS

The above discussion assumed that metagenomic sequencing was performed on nucleic acids extracted from BioWatch filters. Commercial technologies exist that permit targeted sequencing, which can be accomplished via targeted-PCR methods (e.g., pools of PCR primers that amplify genomic regions that are highly discriminating) or via bait-capture methods (e.g., array probes that capture highly discriminating regions, which can be released for library preparation and sequencing once the background is washed away). Targeted sequencing is an option that BioWatch could utilize to look for a specific set of agent targets. Hundreds of regions could be targeted, to provide confident species and strain identification, as well as to identify key genes related to virulence or resistance. The use of a targeted amplification approach could hasten the potential applicability of NGS for BioWatch for confirmation and characterization of a screening PCR result. Deep metagenomic sequencing would still need to be employed in order to detect novel or engineered organisms.

Sequencing as a Replacement for PCR for BioWatch

Given the factors listed above, the committee also asked what requirements would be needed for sequencing to be feasible as a full replacement for PCR for BioWatch. Clearly, the issues of sequencing depth and LOD, total time-to-answer, cost, software to facilitate data analysis, personnel training in NGS, and required information technology and laboratory infrastructure would dominate the calculation. Necessary requirements for BioWatch might include, but not be limited to

- Achieving a pathogen LOD at a level equivalent to current PCR assays;
- Delivering an answer in < 24 hours from sample receipt;
- Costing no more than $1,000/sample (or even a much lower cost that might be roughly comparable to PCR assay costs); and
- Being able to process in batch approximately 36 samples/day (or

some similar number to account for the need to process one or more samples per day from multiple collectors in jurisdictions).

To our knowledge, no existing or announced sequencing technology is able to meet these specifications yet. The DNA sequencing industry currently is driven by the need to lower the cost of sequencing a pure human genome or exome, and the industry has succeeded tremendously in doing so. However, BioWatch faces the need to confidently identify a pathogen at a variable and unknown concentration against a background that can also vary enormously over time (e.g., on high pollen versus low pollen days). This challenge is very different, especially when the required time, cost, and throughput are taken into consideration. Although the technology continues to improve, it is not clear when an appropriate sequencing platform(s) with the necessary performance characteristics will become available. However, the deep analysis of microbiomes, such as those found in the human gut, by NGS is drawing increasing attention from multiple agencies because of the potential health benefits that may be derived. Microbiome analysis may serve as a market driver to produce sequencing platforms optimized in ways that BioWatch can ultimately find useful and leverage. The FDA, CDC, state public health laboratories, and other partners also are exploring the role of high-throughput pathogen genome sequencing using NGS to identify microorganisms in food and environmental samples and compare them with clinical samples from ill patients, in order to identify causative pathogens in outbreaks of foodborne disease. Indeed, approximately 500 isolates of bacteria such as *Salmonella* spp. and *Listeria* spp. are being sequenced each month.[1] These whole-genome sequencing approaches using NGS can also be applied by BioWatch in the future to recover additional genomes of biothreat agents and thereby establish more complete reference databases.

Conclusion About Sequencing Technology for BioWatch

Table 5-1 summarizes ways that three types of next-generation sequencing could be applicable to the BioWatch program, along with current challenges to implementation. In addition to technical challenges, the experience, training and infrastructure within jurisdictional laboratories conducting BioWatch assays will need to be considered.

Ultimately, the committee recommends that the BioWatch program should continue to monitor and evaluate NGS technologies as they develop. Sequencing (through NGS or older Sanger sequencing technol-

[1] Information on the FDA's Whole Genome Sequencing Program is available at http://www.fda.gov/Food/FoodScienceResearch/WholeGenomeSequencingProgramWGS/.

TABLE 5-1 Applications of Next-Generation Sequencing for BioWatch

	Related Terms	Purpose	Advantages	Key Challenges / Barriers to Implementation			
				Turnaround Time	Cost	Complexity of Bioinformatics Analysis	Challenges of Interpretation
Metagenomic sequencing	Shotgun sequencing, random sequencing, unbiased sequencing	Primary BioWatch assay	Most inclusive	X	X	X	X
		Environmental and detection filter background measurement	Enable detection of "natural" positive controls (fungal spores, etc.)		X	X	X
			Better understand sources of false-positive signals				
			Improve design of primers and probe targets used for primary screening				
		Novel pathogen discovery	Identification of novel or divergent pathogens that would go undetected by existing tests		X	X	X
Microbial whole-genome sequencing	Whole-genome resequencing, *de novo* genome assembly	Optimize design of primers and probes	Will identify new targets for PCR and improve primer / probe design		X		
		Expand reference databases	Will better capture target diversity and inform primer / probe design		X	X	
Targeted sequencing	Amplicon sequencing, biased sequencing, conserved region sequencing	Investigate false-positive results	Better understand sources of false-positive signals	X	X		X
		Primary BioWatch assay	More informative than "yes-no" result from real-time PCR; more inclusive of diverse strains	X	X		

ogy) currently could be used to follow up on unexpected assay results from the jurisdictions. In the nearer term, targeted approaches coupled with NGS may be useful to the program as a replacement for the current real-time PCR secondary assays because of their ability to analyze many more genomic regions for identification and characterization. The applicability of metagenomic NGS to the program would be a longer-term prospect.

6 Conclusions and Recommendations

BioWatch is an environmental monitoring system designed to detect genomic material from a set of aerosolized pathogens. It is owned by the Department of Homeland Security (DHS) but relies on a network of program users and partners in state and local jurisdictions as well as federal agencies. The core pathogen detection technology on which BioWatch relies is real-time polymerase chain reaction (PCR) and the laboratory analyses of BioWatch samples are conducted in state and local public health laboratories. To be effective, the program must operate as a public safety and health decision-making tool for federal agencies but, more importantly, for local system users. Responding to a detection that turned out to be false positive drains jurisdictional resources and could undermine confidence in the system when a true positive occurs. On the other hand, a false negative result could be a catastrophic error if a pathogen release was missed.

The BioWatch program drew on scientific knowledge (assays, detection methods, and pathogen genomic organization) that was available at the time of its inception. This base of knowledge has continued to evolve over the course of the program's deployment. Multiple stakeholders, within and outside of DHS and the federal government, have subsequently put substantial effort into the composition of inclusivity, exclusivity, and background environmental panels for microbial detection assays and the development of assay performance and validation standards. In recent years, the BioWatch program also has supported work at Los Alamos National Laboratory (LANL) and Dugway Proving Ground to assemble and qualify strain materials and conduct assay and operational testing of its system's performance. These efforts and existing standards guidance (such as the Stakeholder Panel on Agent Detection Assays [SPADA], Public Health Actionable Assay [PHAA]/Federal Standards for Assay Performance and Equivalency [FSAPE], and from the Food and Drug Administration [FDA] and the Environmental Protection Agency, and others) provide a solid foundation for understanding assay performance and validation standards appropriate for the program's mission.

FUNDAMENTAL PRINCIPLES OF PERFORMANCE STANDARDS AND VALIDATION

An assay performance standard describes requirements that must be met for the assay to be considered acceptable and describes how testing to validate this performance is to be carried out. Elucidation of performance standards that ensure that the BioWatch PCR assays have performance characteristics that provide sufficient confidence to program stakeholders is the motivating subject of this report. Although the report focuses on fundamental principles for the PCR assay component, it is important to emphasize that interpretation of an operational result (which is essential for user "actionability") is a complex process involving multiple components of the system, including sample collection, sample extraction, PCR amplification, PCR detection, data analysis, and current law enforcement or intelligence agency threat assessments. Similarly, validation reflects not only specific testing conducted on the PCR assay's performance, but an overall framework designed to ensure that the quality of the overall systems processes meet their intended purpose.

APPROACHES TO ASSAY PERFORMANCE CHARACTERIZATION AND VALIDATION

BioWatch currently operates using a two-stage testing process. Initial screening assays and reagents are provided through the Department of Defense (DOD) Critical Reagents Program (CRP) (testing for the presence of a single signature per pathogen). Subsequent secondary assays and reagents to verify the screening results are provided through the Centers for Disease Control and Prevention (CDC) Laboratory Response Network (LRN) (testing for the presence of several signatures per pathogen). A screening assay must be able to detect the presence of small amounts of nucleic acid from a tested pathogen (sensitivity) and minimize the rate of obtaining a false negative result. A secondary assay for verification or confirmation, on the other hand, needs to be able to precisely distinguish between the tested pathogen and closely related but nonpathogenic organisms to minimize the rate of a false positive result being passed forward for action (specificity). The combination of the screening and secondary assays thus functions to provide an overall combination of high sensitivity and high specificity.

The BioWatch program operates under some historical assumptions and program constraints. These include using one screening assay per BioWatch threat organism, emphasizing stringent performance of the initial assay (as noted above, certainly such assays must not produce many false negatives), using a secondary PCR panel of 3-5 target signatures, and not incorporating additional rule-out assays for certain agents. These assumptions are challenged by the strain variability in agents of interest to BioWatch. Newly sequenced strains of target agents eventually will reveal potential false negative situations, and potential false positive situations from newly discovered near-neighbor genomes. Similarly, the program has focused its consideration of the development of multiplex assays primarily on the initial screen, which may reflect earlier days of multiplex PCR development, when successfully multiplexing more than a handful of assays was more difficult. Current PCR multiplexing, in solution or solid phase formats, is no longer subject to as many constraints, and BioWatch now can consider expanded multiplexed approaches that incorporate a larger number of assays to determine presence or absence of key threat agents.

A final assumption appears to be that the primary opportunity for significant assay performance testing against large inclusivity, exclusivity, and environmental panels is during laboratory validation. The biothreat agent releases the BioWatch system is designed to detect are rare events, against complex environmental backgrounds, and all assays have associated false positive and false negative rates. Using laboratory assay testing to obtain very high levels of statistical confidence that might be desirable for decision making (e.g., that an assay hypothetically be shown to have a false positive or false negative error rate of less than 1 in 100,000 tests with 95 percent confidence) is not practically achievable:

- The ongoing analysis of assay performance data from verification in user laboratories and from operational deployment of the assay in the program is critical.
- It will not be possible to undertake laboratory validation testing using sufficient replicates on enough different inclusivity, exclusivity, and environmental organisms in the presence of enough different types of potentially interfering substances and different filter conditions to characterize the long-term false positive and false negative rates of an assay to a level of statistical confidence users may desire operationally.

A process for testing assay performance to some reasonable level in the laboratory, balancing time, costs, and other constraints, is needed in concert with ongoing analysis of data obtained from assay verification in user laboratories and from operational deployment of the assay in the program. These additional sources of data will be important in providing further information on assay performance in jurisdictional settings and in obtaining a more accurate sense of assay false positive and false negative rates. This process will play a role in building confidence in the assay.

The use of rapidly expanding sequence databases can be of use to the program, and the increasing use of massively multiplexed assays also is changing the nature of some validation strategies. For example, when an individual assay in a multiplex is modified or replaced to reflect new knowledge of circulating strains, or a new assay is added to the multiplex, a subset of validation testing can be conducted rather than repeating the full range of performance data.

In addition to giving further consideration to the possibilities enabled by the use of multiplex assays, the report suggests several options that BioWatch could consider as alternatives to the model of initial screening and subsequent secondary assays and the associated processes of performance characterization and validation. These include

- Expanding the use of *in silico* screening during assay characterization and validation;
- Using a tiered or serial testing approach with certain performance characteristics validated more stringently than others based on the intended use of the assay;
- Developing subsets of assays for each of the tested BioWatch agents rather than a single set of screening and secondary assays (e.g., including additional types of rule-in and rule-out assays along with decision algorithms on when to use which tests); and
- Incorporating next-generation sequencing (NGS) into the program as technology and software development makes this increasingly feasible.

Recognizing the Work That Has Been Done

It is important to recognize the substantial body of work that the BioWatch program and its contractors have undertaken to assemble reference materials, design appropriate testing setups, provide guidance on performance standards, test and evaluate PCR assays, conduct testing to understand current system performance, and establish and operate a quality assurance (QA) program. A significant base of information characterizing assay and system performance is available now to support the program's successful operations.

A PCR Assay Standard to Meet BioWatch's Mission

The committee focused on the BioWatch program's mission in order to explore the technical considerations that inform design of appropriate PCR performance standards for the program and how they could be optimized. To be applicable to BioWatch, a PCR standard to characterize assay performance and conduct validation would need to meet the following basic goals:

- Establish that the assay has sufficient sensitivity to detect the release of a tested biothreat agent at a program-relevant concentration above the baseline of environmental background;
- Establish that the assay has sufficient specificity to detect tested pathogenic strains of concern but not to cross-react with non-tested strains and organisms within acceptable program false positive and false negative rates; and
- Be sufficiently robust for routine operational use by BioWatch jurisdictional laboratories and be otherwise acceptable to program users.

The committee's conclusions and recommendations to address these goals are detailed below.

Testing Strategies: Opportunities to Expand *in Silico* Testing

The committee sees opportunities for BioWatch to increase its use of *in silico* testing. To make such testing successful, parallel efforts will be needed to ensure that reference databases contain as many high-quality sequences as possible (Chain et al., 2009) and are available to PCR assay and device developers through accessible reference databases. The availability of high-quality reference databases also will support PCR assay design by enabling signatures, primers, and probes to precisely target inclusivity strains while avoiding cross-reactivity with exclusivity strains (within the limits achievable given microbial diversity).

Recommendations

BioWatch should strive to test its PCR assays against as many known strains of its target agents and relevant near neighbors as possible. Undertaking all such testing through laboratory analysis alone is likely to be cost, time, and feasibility prohibi-

tive. Increased *in silico* testing should thus be incorporated into the program to support targeted laboratory testing.

To support this goal, the BioWatch program should work with other federal agencies and international partners to ensure that as many strains as possible are sequenced to at least a high-quality draft level and the data and associated reference materials made available to PCR assay and device developers through reference databases.

Assay Sensitivity and Specificity

Assay sensitivity relates to the ability of the assay to reliably detect the targeted organism at very low concentrations. Assay specificity is the ability to reliably detect all variants of the targeted organism without detecting nontargeted near-neighbor organisms. For pathogen detection assays such as those used by BioWatch, characterization of these parameters (whether defined in precisely this manner or defined under the overall probability-of-detection [POD] concept articulated by AOAC) provides important information for decision makers. The committee concluded the following:

- Either approach—using a predetermined acceptable minimum detectable level (AMDL), if there is stakeholder agreement on this value, or calculating minimum concentration of nucleic acid detectable by the assay with sufficient confidence, followed by stakeholder determination of whether this makes the assay acceptable for the mission—is reasonable to establish assay sensitivity as part of a performance standard.
- Using a reference strain approach to determining the limit of detection (LOD) and POD is reasonable given practical constraints on testing, but variation in true LOD/POD among strains is a potential issue of which programs making operational use of such assays should be aware and document. A reasonable option to help address this situation is to test a limited number of reference strains (such as n = 3-5) in a range of quantities near the expected LOD to get a sense of variation.

Similarly, the committee makes the following recommendations regarding inclusivity and exclusivity strains:

Recommendations

The panels of inclusivity and exclusivity strains specified for laboratory testing should seek to balance broad phylogenetic

and geographic diversity within the realm of practical testing. Inclusivity panel strains for laboratory testing should sample the genetic diversity represented by available strain collections with an emphasis on prioritizing those strains that cause significant morbidity and mortality, have high transmissibility, have a wide host range, or may be readily accessed by nefarious actors.

For laboratory testing of exclusivity panels (near neighbors and environmental organisms), a strategy of sample pooling is appropriate to increase testing efficiency. If a positive result is detected, each agent in the pool should be tested separately.

The PCR is Only One Step in the Process of a PCR Assay

The current study and resultant report have focused primarily on the core PCR reaction portion of a complete PCR assay, and the key processes before and after the PCR are not significantly addressed. PCR assay performance is critically impacted by the steps that lead into the assay and is influenced by the amplicon detection methodology and feedback from the steps that follow a result. Even a perfectly designed PCR with ideal specificity and sensitivity will fail with either a false positive or false negative, or an inconclusive, result if the sample collection, processing, purification, and reagent input are not optimized and standardized. In addition, detection methodology requires certain characteristics of the amplicon that results from the reaction. The ability to securely identify the target of interest can be affected by these interacting factors. Even the comprehensive work performed at LANL to establish standards, which was done with purified DNA samples, experienced issues with reagents. Similar rigorous and comprehensive studies should be performed, developed, and implemented for BioWatch in terms of sample processing and amplicon detection.

Recommendation

A PCR assay performance standard that meets program and user needs should include development and implementation of standards that encompass all of the individual steps in the assay from sample collection, extraction, amplification, detection, reagent quality control, and data analysis. There should be an integrated assay standard, as well, that includes all of the steps of a complete PCR assay as a unit.

Environmental Background

The BioWatch system is required to successfully detect a set of pathogenic microorganisms from a multitude of nonpathogenic or unrelated microorganisms present in the environment. This feature requires understanding how the assay result differs from a baseline. How to detect a true signal from background noise is complicated by the fact that the distribution of bacteria and viruses in different environments around the United States is not well characterized. The nucleic acid extracted from BioWatch collection filters must be amplified and detected from the complex background of environmental substances (e.g., soils) that may interfere with the assay. The committee drew the following conclusions:

- The information obtained from knowledge of environmental background could be used to prioritize the strains included in the environmental panel specified by the performance standard, in order to cover genetic diversity while keeping the panel size manageable. The information also could be used to improve the interpretation of assay results in the context of individual jurisdictions and minimize potential environmental nonthreat detections, and to provide a basis for improved understanding of assay predictive value in a particular jurisdiction or collector location.
- Expanded *in silico* testing is not likely to reduce the burden of laboratory testing against potentially interfering environmental substances. Conducting performance testing against a panel of such substances is useful to provide information on possible performance issues and to ensure sufficient assay performance in an operational context.

Recommendations

The Department of Homeland Security should undertake further characterization of environmental background organisms over time and at various locations. This sampling could build on the existing BioWatch network by using previously tested filters as samples and should also include judicious sampling of water and soil samples. Operational assay performance data in individual jurisdictions based on positive screening results might provide a source of invaluable data, as well.

Assay performance in the presence of known environmental contamination (e.g., rail dust in a subway system) should be addressed by assessing the assay detection limit for a limited number of inclusivity and exclusivity panel agents spiked on jurisdictional background filters.

A systematic process should be instituted to identify the root cause of positive assay results that do not appear to represent the targeted microorganism and to determine appropriate corrective actions to address the cause(s). This feedback and revision process should include the participation of laboratory experts in the jurisdictions and federal experts.

Reference Materials

A source of reference materials and reagents that is accessible to those who need it for development and validation is critical to the application of a performance standard. The DOD CRP currently serves this function for the screening assay and for quality control samples tested under the BioWatch QA program. The secondary assay and associated reagents are provided through the CDC LRN. The committee concluded that

- Maintaining and quality controlling reference materials (particularly materials derived from microbial strain cultures) are not trivial or inexpensive processes. Not all strains of a pathogen reported in the literature or identified clinically are easily accessible. Furthermore, the genomic organization of pathogens maintained in culture for extended time periods may no longer accurately reflect that of their pathogenic predecessors and some data simply are incorrect.
- The CRP credits its predicable and sustained funding through DOD as a necessary element for successful development and maintenance of its reference materials. If some or all of the materials needed for BioWatch assay performance testing and validation (now or in the future) will not be provided through CRP, a similar type of quality-controlled, sustainable, and accessible repository will need to be developed independently by DHS.

Recommendation

Federal agencies, including the Department of Homeland Security, the Department of Defense, the Department of Energy, the Department of Health and Human Services, and other partners, should collaborate to produce a reference document or database identifying which repositories contain strains or strain nucleic acids relevant for testing pathogen detection assays. Information on the requirements or procedures to obtain and use strains or strain materials (e.g., extracted nucleic acids) held by these repositories should be included, where possible. This information would provide a coherent source of information to

both federal agency programs and to individuals or companies involved in assay and device development and validation.

The Key Importance of Ongoing Assay and Reference Database Maintenance

Periodically re-examining performance is part of the ongoing process of QA. The performance of pathogen identification assays are inherently a process and not an end point: new strains and near neighbors will always be encountered, and the microorganisms themselves are constantly evolving. New sequence information continues to be obtained, which results in a dwindling number of potential unique PCR assays available for species/strain differentiation. A corollary is that for some agents of potential interest to BioWatch, it may not be possible to uniquely specify appropriate targets because of closely overlapping genomes among the target and nonpathogenic near neighbors. There already are signs of this with some strains of *Bacillus cereus* that are lethal to humans, very similar to *Bacillus anthracis*.

- BioWatch should be prepared to flexibly alter the PCR assays to accommodate these kinds of circumstances. If not, the next set of BioWatch assays is just moving the boundaries of the Maginot Line and will likely gradually but inexorably evolve toward obsolescence.
- The program could also consider including an additional, complementary agent test to achieve maximum sensitivity and specificity in its detection capabilities for certain agents, or for use in certain jurisdictions experiencing issues with a specific assay.

Similarly, documentation of the limits of the assay and well-described interpretation criteria provide important information so that users do not overextend their use of the data. A limitation of making knowledge actionable is not dependent on having perfect data, but on understanding the assumptions on and limitations of the data in order to effectively interpret results.

- When providing or deploying an assay, information on any potential issues with the assay (e.g., that strains *x* or *y* have been shown to cross-react or that the limit of detection is decreased to *xx* in the presence of contaminant *yy*) needs to be provided and built into operational guidance. Regularly updating the assay with this type of information as it becomes available may be easier than going back through full redesign and revalidation of the assay if it can enable appropriate decision making based on the assay in an operational setting.

Recommendation

The BioWatch program should institute a process to regularly review and update all assays in light of new genetic knowledge. The program should re-evaluate its assays through at least annual *in silico* screening to take advantage of new genomic information in databases, update signatures, and identify potential false positive or false negative issues. Similarly, the composition of inclusivity, exclusivity, and environmental panels used for testing should be reviewed regularly and laboratory re-testing conducted as needed. Annual *in silico* and periodic laboratory re-testing of assay performance should be part of any assay performance standard accepted for BioWatch.

Starting Point for a BioWatch Assay Performance Standard

As discussed in the report, some reasonable compromise must be achieved in order to undertake performance characterization and validation work on assays deployed or being considered for deployment in the program. The committee has attempted to provide a starting point for a performance standards approach for conducting laboratory testing of assays that could provide reasonable statistical confidence for BioWatch users while seeking to keep numbers of tests and numbers of strains reasonable for intended use.

Recommendation

The following approach can serve as a starting point for a standard to provide confidence in PCR performance while seeking to undertake a reasonable amount of laboratory characterization and validation testing:

- **A reasonable approach to determining an assay's analytical LOD in a laboratory setting is to conduct serial dilution at a range of concentrations bracketing the estimated LOD, using $n = 60$ replicates with acceptance criteria of at least 58/60 for a given concentration, followed by appropriate curve fitting. LOD testing should also be conducted in the presence of realistic background matrix, such as previously tested BioWatch filters, to gain an understanding of how the LOD may be affected by operational background.**
- **The use of *in silico* screening presents a relatively fast and low-cost way to predict one part of assay performance using**

as many strains as are available as high-quality sequences in reference databases. *In silico* screening should be undertaken using the set of PHAA panels, if these strains' sequences are available.

- **The strain panels recommended by SPADA represent a good starting point for laboratory validation testing of BioWatch assays and were developed with stakeholder participation. However, many were developed in 2010-2011 and should be reviewed and updated to account for new genetic knowledge.**
- **The strategy used by LANL for testing detectability or non-detectability of inclusivity and exclusivity strains represents a reasonable model for BioWatch (testing n = 20 replicates, followed by testing 20 further replicates if an unexpected detect or nondetect result is obtained, with acceptance criteria of 20/20 or 39/40).**

A ROBUST QUALITY ASSURANCE PROGRAM CONTRIBUTES TO CONFIDENCE IN THE SYSTEM BUT COMMUNICATION CHALLENGES REMAIN

Process, Transparency, and Communication

Various technical factors should be considered when establishing assay performance standards appropriate for BioWatch's mission. The report discusses key principles and provides the committee's conclusions and recommendations in these areas. Equally important, however, is the process that is used to establish such standards.

- Different performance standards have taken different approaches in developing their guidance. Although the BioWatch program is federally owned and relies on important federal relationships, it cannot succeed without close engagement with nonfederal partners. The use of subject-matter experts and inclusion of a stakeholder forum, of whatever type, in coming to agreement on the performance and validation standards used by the program will be key to ensuring that the standards are acceptable and meet program and user needs.

Since 2011, BioWatch has implemented a QA program that helps support the continuous process of validation, provides a mechanism to continue to build relationships with BioWatch jurisdictional laboratories, and provides a source of operational performance data that can be used by the program to improve the understanding and interpretation of assay results. The BioWatch program also has made important strides since program inception in improving communication with state and local jurisdictions. Despite this progress,

- Jurisdictions, particularly jurisdictional laboratory personnel, indicate that they do not have access to the full range of assay performance validation and operational performance data they would like in order to support decision making based on assay results. This situation applies to both the initial screening assay and the secondary assay. This discrepancy limits the performance knowledge of those responsible for making response recommendations based on that very knowledge. As a result, there remains a lack of confidence in the assays from users in the jurisdictions. The committee does not lay blame on any one group for the communication deficiency. However, a system of effective data sharing and communication is a critical component of overall program success. There is an essential need for additional dialogue to implement a data-sharing and response system that meets program office, federal partner, and jurisdictional needs.
- Communication also may be useful in establishing which sample extraction and preparation methods work most effectively under which ranges of circumstances, and in helping to establish greater commonality in PCR input steps among jurisdictions.

Recommendation

Communication and information-sharing are necessary to establish confidence in a system. The BioWatch program, relevant federal agencies, and local and state jurisdictions should expand the communication and data-sharing that occurs among the network of federal and nonfederal partners involving both the screening and the secondary PCR assays. This effort would assist in establishing acceptable performance standards, enhance informed data interpretation and decision making, improve the ability to undertake root cause analysis of assay issues encountered by jurisdictions, and enable the collective identification and dissemination of actions as part of robust quality assurance.

STANDARDS APPROACHES FOR MULTIPLEXING REAL-TIME PCR ASSAYS

The committee concluded that existing standards approaches should be suitable for characterizing and validating the performance of multiplex PCR assays. In particular, guidance recently generated by the FDA on highly multiplexed PCR assays provides information on the types of bridging studies or targeted validation studies that can be undertaken when making an alteration to an existing multiplex assay. In this way, all studies undertaken during the original multiplex assay validation do not need to be fully repeated, striking a balance between the need for updated performance data and practical time and cost constraints.

Additional considerations will arise with multiplex assays compared to singleplex assay. These include the complications posed by the potential for probe cross-hybridization, which decreases specificity, and practical challenges with validating mixtures of targets versus a single target. The ability for developers to successfully multiplex assays continues to improve, and there are opportunities for the BioWatch program to move in the direction of multiplexed assays. A critical issue with the prior deployment of multiplex assays in the program was a lack of user confidence in assay performance, compounded by a lack of available performance data. Expanded data sharing (previous recommendation) should help to address this concern and enable stakeholders to reach agreement on whether particular multiplex agent detection assays are acceptable for use in BioWatch.

The committee concluded that existing guidance on singleplex PCR performance for environmental biodetection assays in concert with FDA information should provide a good starting point for the performance testing and validation of multiplex assays by the BioWatch program. The additional issues that arise when designing, validating, and deploying multiplex assays will need to be considered when making the decision to switch to a multiplex format.

INCORPORATING ADVANCES IN SEQUENCING TECHNOLOGY INTO THE PROGRAM

Rapid developments in technology such as NGS have significant implications for the program. The program eventually may be able to move away from the limitations of PCR assays that target only a few small regions of a microbe to a more definitive form of agent identification that relies on information recovered from an expanded portion of the genome. Metagenomic NGS, in particular, is attractive in the future as a universal, target-independent approach to identify novel infectious agents, including emerging pathogens associated with outbreaks (Briese et al. 2009; Grard et al. 2012).

In the short term, the program can use sequencing as a tool to help resolve unusual assay results. In this way, sequencing could function as a tertiary confirmation and characterization tool. Sequencing and associated informatics are not yet at a stage in which they could replace PCR assays as the program's principal detection methodology, and many challenges remain in detecting a low-prevalence target sequence against a high-complexity environmental background in a reasonable period of time. However, the program should monitor developments in this area as speed increases, cost decreases, and required informatics and databases improve.

Recommendation

The Department of Homeland Security should monitor developments in technology, such as next-generation sequencing, as speed increases, cost decreases, and required informatics and databases improve. A program to evaluate technology and incorporate it into BioWatch would enable the program to improve its ability to address current challenges and enable it to meet those of the future. However, the program should plan to work with at least some early adopter laboratory users in the BioWatch jurisdictions, along with technology experts, to make sure that new technology brought into the program not only incorporates the best technology but also functions smoothly for the stakeholder community.

BioWatch is only one tool in a layered network of public health and safety disease surveillance activities. The challenges that the program faces are complex, and to be successful the results obtained from the system's collectors must be useable by decision makers in state and local jurisdictions. Currently, and for at least the near future, the program relies on real-time PCR as the core of its agent detection technology. The program has a strong foundation from which it can develop assay performance and validation standards that meet program needs. As the report observes, however, the most striking issues do not appear to be technical challenges. Rather, they are primarily communication and education and training. Building additional, specific mechanisms into the program to share assay and system performance results, discussing what these performance results reveal about the limits of the assay and the translation of assay results into actions, and sharing applicable assay performance information across agencies such as DHS and CDC and with appropriate individuals within the network of jurisdictions appear to the committee to be critical needs. Even while recognizing the necessary security constraints, this goal should be achievable. Addressing this issue would go far in ensuring confidence in the system and its results in order to facilitate responses to a BioWatch Actionable Result.

References

Anthony, S.J., J.H. Epstein, K.A. Murray, I. Navarrete-Macias, C.M. Zambrana-Torrelio, A. Solovyov, R. Ojeda-Flores, N.C. Arrigo, A. Islam, S. Ali Khan, P. Hosseini, T.L. Bogich, K.J. Olival, M.D. Sanchez-Leon, W.B. Karesh, T. Goldstein, S.P. Luby, S.S. Morse, J.A. Mazet, P. Daszak, and W.I. Lipkin. 2013. A strategy to estimate unknown viral diversity in mammals. *mBio* 4(5):e00598-13. Available: http://mbio.asm.org/content/4/5/e00598-13 [accessed December 2, 2014].

AOAC (AOAC International). 2011a. SMPR-2010.001, Standard method performance requirements for polymerase chain reaction (PCR) methods for detection of *Francisella tularensis* in aerosol collection filters and/or liquids. *Journal of AOAC International* 94(4):1338-1341.

AOAC. 2011b. SMPR-2010.002, Standard method performance requirements for polymerase chain reaction (PCR) methods for detection of *Yersinia pestis* in aerosol collection filters and/or liquids. *Journal of AOAC International* 94(4):1342-1346.

AOAC. 2011c. SMPR-2010.003, Standard method performance requirements for polymerase chain reaction (PCR) methods for detection of *Bacillus anthracis* in aerosol collection filters and/or liquids. *Journal of AOAC International* 94(4):1347-1351.

AOAC. 2011d. SMPR-2010.004, Standard method performance requirements for immunological-based handheld assays (HHAs) for detection of *Bacillus anthracis* spores in visible powders. *Journal of AOAC International* 94(4):1352-1355.

AOAC. 2011e. SMPR-2009.005, Standard method performance requirements for immunological-based handheld assays (HHAs) for detection of ricin in visible powders. *Journal of AOAC International* 94(4):1356-1358.

AOAC 2014. SMPR-2014.006, Standard method performance requirements for detection and identification of Variola virus DNA in aerosol collection filters and/or liquids. Approved by AOAC Stakeholder Panel on Agent Detection Assays (SPADA). Final Version Date: August 14, 2014. Effective Date: October 15, 2014.

AOAC. In preparation-a. SMPR-2011.xxx, Standard method performance requirements for polymerase chain reaction (PCR) methods for detection of *Burkholderia mallei* in aerosol collection filters and/or liquids.

AOAC. In preparation-b. SMPR-2011.xxx, Standard method performance requirements for polymerase chain reaction (PCR) methods for detection of *Burkholderia pseudomallei* in aerosol collection filters and/or liquids.

AOAC Methods Committee. 2011. Methods Committee guidelines for validation of biological threat agent methods and/or procedures. *Journal of AOAC International* 94(4):1359-1381.

Barns, S.M., C.C. Grow, R.T. Okinaka, P. Keim, and C.R. Kuske. 2005. Detection of diverse new *Francisella*-like bacteria in environmental samples. *Applied and Environmental Microbiology* 71(9):5494-5500.

Bean, W.J. Jr., N.J. Cox, and A.P. Kendal. 1980. Recombination of human influenza A viruses in nature. *Nature*. 284(5757):638-640.

Briese, T., J.T. Paweska, L.K. McMullan, S.K. Hutchison, C. Street, G. Palacios, M.L. Khristova, J. Weyer, R. Swanepoel, M. Egholm, S.T. Nichol, and W.I. Lipkin. 2009. Genetic detection and characterization of Lujo virus, a new hemorrhagic fever-associated arenavirus from southern Africa. *PLoS Pathogens* 5(5):e1000455. Available: http://www.plospathogens.org/article/info%3Adoi%2F10.1371%2Fjournal.ppat.1000455 [accessed December 3, 2014].

Brown, L.D., T.T. Cai, and A. DasGupta. 2001. Interval estimation for a bionomial proportion. *Statistical Science* 16(2):101-133.

Budowle, B., S.E. Schutzer, S.A. Morse, K.F. Martinez, R. Chakraborty, B.L. Marrone, S.L. Messenger, R.S. Murch, P.J. Jackson, P. Williamson, R. Harmon, and S.P Velsko. 2008. Criteria for validation of methods in microbial forensics. *Applied Environmental Microbiology* 74(18):5599-5607.

Budowle, B., N.D. Connell, A. Bielecka-Oder, R.R. Colwell, C.R. Corbett, J. Fletcher, M. Forsman, D.R. Kadavy, A. Markotic, S.A. Morse, R.S. Murch, A. Sajantila, S.E. Schmedes, K.L. Ternus, S.D. Turner, and S. Minot. 2014. Validation of high throughput sequencing and microbial forensics applications. *Investigative Genetics* 5:9 [online]. Available: http://www.investigativegenetics.com/content/5/1/9 [accessed November 18, 2014].

Burd, E.M. 2010. Validation of laboratory-developed molecular assays for infectious diseases. *Clinical Microbiology Reviews* 23(3):550-576.

Bush, G.W. 2003. State of the Union Address. Washington Post, January 28 [online]. Available: http://www.washingtonpost.com/wp-srv/onpolitics/transcripts/bushtext_012803.html [accessed November 17, 2014].

Bustin, S.A., V. Benes, J.A. Garson, J. Hellemans, J. Huggett, M. Kubista, R. Mueller, T. Nolan, M.W. Pfaffl, G.L. Shipley, J. Vandesompele, and C.T. Wittwer. 2009. The MIQE guidelines: Minimum information for publication of quantitative real-time PCR experiments. *Clinical Chemistry* 55:611-622.

Case, R.J., Y. Boucher, I. Dahllöf, C. Holmström, W.F. Doolittle, and S. Kjelleberg. 2007. Use of 16S rRNA and rpoB genes as molecular markers for microbial ecology studies. *Applied and Environmental Microbiology* 73(1):278-288.

Chain, P.S.G., D.V. Grafham, R.S. Fulton, M.G. FitzGerald, J. Hostetler, D. Muzny, J. Ali, B. Birren, D.C. Bruce, C. Buhay, J.R. Cole, Y. Ding, S. Dugan, D. Field, G.M. Garrity, R. Gibbs, T. Graves, C.S. Han, S.H. Harrison, S. Highlander, P. Hugenholtz, H.M. Khouri, C.D. Kodira, E. Kolker, N.C. Kyrpides, D. Lang, A. Lapidus, S.A. Malfatti, V. Markowitz, T. Metha, K.E. Nelson, J. Parkhill, S. Pitluck, X. Qin, T.D. Read, J. Schmutz, S. Sozhamannan, P. Sterk, R.L. Strausberg, G. Sutton, N.R. Thomson, J.M. Tiedje, G. Weinstock, A. Wollam, Genomic Standards Consortium Human Microbiome Project Jumpstart Consortium, and J.C. Detter. 2009. Genome project standards in a new era of sequencing. *Science* 326(5950):236-237.

Chiu, C.Y. 2013. Viral pathogen discovery. *Current Opinion in Microbiology* 16(4):468-478.

CLSI (Clinical and Laboratory Standards Institute; formerly National Committee on Clinical Laboratory Standards). 2004. Protocols for Determination of Limits of Detection and Limits of Quantitation; Approved Guideline. NCCLS Document EP17-A. Wayne, PA: NCCLS.

CLSI. 2012. Evaluation of Detection Capability for Clinical Laboratory Measurement Procedures; Approved Guideline, 2nd ed. CLSI document EP17-A2. Wayne, PA: Clinical and Laboratory Standards Institute.

Coates, S.G., S.L. Brunelle, and M.G. Davenport. 2011. Development of standard method performance requirements for biological threat agent detection methods. *Journal of AOAC International* 94(4):1328-1337.

CRP (Critical Reagents Program). 2015. *Critical Reagents Program FY15 Product Catalog*, Version 1. Available: www.jpeocbd.osd.mil/packs/DocHandler.ashx?DocID=14895 [accessed January 15, 2015].

Davenport, M.G. 2014. Stakeholder Panel on Agent Detection Assays. Presentation to the Committee on PCR Standards for the BioWatch Program, March 13, Washington, DC.

DOD (Department of Defense). 2005. Department of Defense Chemical and Biological Defense Program Annual Report to Congress. March.

Domingo, E., J. Sheldon, and C. Perales. 2012. Viral quasispecies evolution. *Microbiology and Molecular Biology Reviews* 76(2):159-216.

Dorai-Raj, S. 2014. Binomial confidence intervals for several parameterizations. Version 1.1-1. Package "binom" for R. Available: http://cran.r-project.org/web/packages/binom/binom.pdf [accessed January 9, 2015].

DHS (Department of Homeland Security). 2011. *Implementation Plan for Federal Standards for Evaluating Assay Performance and Equivalency in Support of National Biomonitoring Architecture—Nucleic Acid Based Detection Assays*. August 26..

DHS. 2013a. *Public Health Actionable Assays (PHAA) Standards of Implementation Independent Statistical Assessment*. BioWatch Quality Assurance Program. October 2.

DHS. 2013b. SPADA Standard Method Performance Requirements Independent Statistical Assessment. BioWatch Quality Assurance Program, October 2.

DHS. 2014. Government Unique Standards for Implementation of Public Health Actionable Assays—Nucleic Acid Based Diagnostic/Detection Assays (Clinical and Non-Clinical Samples). May 27.

Emanuel, P. 2014. Presentation to the Committee on PCR Standards for the BioWatch Program, June 12, Washington, DC.

EPA (Environmental Protection Agency). 2009. Method Validation of U.S. Environmental Protection Agency Microbiological Methods of Analysis. Prepared by the FEM Microbiology Action Team for the EPA Forum on Environmental Measurements (FEM). FEM Document Number 2009-01. October 7.

FDA (Food and Drug Administration). 2008. FDA clears first test designed to detect and identify 12 respiratory viruses from single sample. FDA News Release. January 3 Available: http://www.fda.gov/NewsEvents/Newsroom/PressAnnouncements/2008/ucm116829.htm [accessed November 11, 2014].

FDA. 2011. Guidelines for the Validation of Analytical Methods for the Detection of Microbial Pathogens in Foods. FDA Methods Validation Guidelines for Microbial Pathogens, Office of Foods. Available: http://www.fda.gov/ScienceResearch/FieldScience/ucm273423.htm [accessed November 18, 2014].

FDA. 2014a. Highly Multiplexed Microbiological/Medical Countermeasure In Vitro Nucleic Acid Based Diagnostic Devices—Guidance for Industry and Food and Drug Administration Staff. Document. August 27. Available: http://www.fda.gov/MedicalDevices/DeviceRegulationandGuidance/GuidanceDocuments/ucm327293.htm [accessed November 11, 2014].

FDA. 2014b. Letter to Robert E. Miller, U.S. Department of Defense, from Margaret E. Hamburg, Commissioner of Food and Drugs, October 10, 2014. Available: http://www.fda.gov/downloads/MedicalDevices/Safety/EmergencySituations/UCM418799.pdf [accessed November 11, 2014].

FDA. 2014c. Quality system regulation. *Code of Federal Regulations*, Title 21, Volume 8, Part 820, Section 3. Available: http://www.accessdata.fda.gov/scripts/cdrh/cfdocs/cfcfr/CFRSearch.cfm?fr=820.3 [accessed October 31, 2014].

GAO (Government Accountability Office). 2009. Biosurveillance: Developing a Collaboration Strategy Is Essential to Fostering Interagency Data and Resource Sharing. GAO-10-171. Washington, DC: GAO. Available: http://www.gao.gov/assets/300/299667.pdf [accessed October 28, 2014].

GAO. 2010. Biosurveillance: Efforts to Develop a National Biosurveillance Capability Need a National Strategy and a Designated Leader. GAO-10-645. Washington, DC: GAO. Available: http://www.gao.gov/assets/310/306362.pdf [accessed October 28, 2014].

GAO. 2011. Biosurveillance: Nonfederal Capabilities Should Be Considered in Creating a National Biosurveillance Strategy. GAO-12-55. Washington, DC: GAO. Available: http://www.gao.gov/assets/590/586047.pdf [accessed October 28, 2014].

GAO. 2012a. Biosurveillance: DHS Should Reevaluate Mission Need and Alternatives Before Proceeding with BioWatch Generation-3 Acquisition. GAO-12-810. Washington, DC: GAO. Available: http://www.gao.gov/assets/650/648026.pdf [accessed October 28, 2014].

GAO. 2012b. Chemical, Biological, Radiological, and Nuclear Risk Assessments: DHS Should Establish More Specific Guidance for Their Use. GAO-12-272. Washington, DC: GAO. Available: http://www.gao.gov/assets/590/587674.pdf [accessed January 6, 2015].

Garza, A. 2012. *The truth about BioWatch: The importance of early detection of a potential biological attack.* June 12, 2012. Washington, DC. Available: http://www.dhs.gov/blog/2012/07/12/truth-about-biowatch [accessed October 29, 2014].

Gire, S.K., A. Goba, K.G. Andersen, R.S. Sealfon, D.J. Park, L. Kanneh, S. Jalloh, M. Momoh, M. Fullah, G. Dudas, S. Wohl, L.M. Moses, N.L. Yozwiak, S. Winnicki, C.B. Matranga, C.M. Malboeuf, J. Qu, A.D. Gladden, S.F. Schaffner, X. Yang, P.P. Jiang, M. Nekoui, A. Colubri, M.R. Coomber, M. Fonnie, A. Moigboi, M. Gbakie, F.K. Kamara, V. Tucker, E. Konuwa, S. Saffa, J. Sellu, A.A. Jalloh, A. Kovoma, J. Koninga, I. Mustapha, K. Kargbo, M. Foday, M. Yillah, F. Kanneh, W. Robert, J.L. Massally, S.B. Chapman, J. Bochicchio, C. Murphy, C. Nusbaum, S. Young, B.W. Birren, D.S. Grant, J.S. Scheiffelin, E.S. Lander, C. Happi, S.M. Gevao, A. Gnirke, A. Rambaut, R.F. Garry, S.H. Khan, and P.C. Sabeti. 2014. Genomic surveillance elucidates Ebola virus origin and transmission during the 2014 outbreak. *Science* 345(6202):1369-1372.

Grard, G., J.N. Fair, D. Lee, E. Slikas, I. Steffen, J.J. Muyembe, T. Sittler, N. Veeraraghavan, J.G. Ruby, C. Wang, M. Makuwa, P. Mulembakani, R.B. Tesh, J. Mazet, A.W. Rimoin, T. Taylor, B.S. Schneider, G. Simmons, E. Delwart, N.D. Wolfe, C.Y. Chiu, and E.M. Leroy. 2012. A novel rhabdovirus associated with acute hemorrhagic fever in central Africa. *PLoS Pathogens* 8(9):e1002924. Available: http://www.plospathogens.org/article/info%3Adoi%2F10.1371%2Fjournal.ppat.1002924

Goodwin, B. 2014. Presentation to the Committee on PCR Standards for the BioWatch Program, June 12, Washington, DC.

Hobson, J.P. 2014. Presentation to the Committee on PCR Standards for the BioWatch Program, June 12, Washington, DC.

Hospodsky, D., N. Yamamoto, and J. Peccia. 2010. Accuracy, precision, and method detection limits of quantitative PCR for airborne bacteria and fungi. *Applied and Environmental Microbiology* 76(21):7004-7012.

IOM and NRC (Institute of Medicine and National Research Council). 2011. *BioWatch and Public Health Surveillance: Evaluating Systems for the Early Detection of Biological Threats. Abbreviated Version.* Washington, DC: The National Academies Press.

IOM and NRC. 2014. *Technologies to Enable Autonomous Detection for BioWatch: Ensuring Timely and Accurate Information for Public Health Officials: Workshop Summary.* Washington, DC: The National Academies Press.

Isbell, M. 2014. PCR standards for the BioWatch Program: BioWatch and statistical assessment of implementation. Presentation to the Committee on PCR Standards for the BioWatch Program, June 12, Washington, DC.

Jennings, L., V.M. Van Deerlin, M.L. Gulley, and College of American Pathologists Molecular Pathology Committee. 2009. Recommended principles and practices for validating clinical molecular pathology tests. *Archives of Pathology and Laboratory Medicine* 133:743-755.

Keim, P. 2014. *SPADA's Burkholderia Working Group Report*. Presentation to the Committee on PCR Standards for the BioWatch Program, June 12, Washington, DC.

Kuske, C.R., S.M. Barns, C.C. Grow, L. Merrill, and J. Dunbar. 2006. Environmental survey for four pathogenic bacteria and closely related species using phylogenetic and functional genes. *Journal of Forensic Sciences* 51(3):548-558.

Lindler, L. 2014. Selection of *Yersinia pestis* assay validation panels. Presentation to the Committee on PCR Standards for the BioWatch Program, June 12, Washington, DC.

LANL (Los Alamos National Laboratory). 2010. Test Plan for the Evaluation of Generation 3 Vendor Assays. April 8. Department of Homeland Security (DHS) BioWatch Project. Los Alamos, NM: Los Alamos National Laboratory.

LANL. 2011a. Assembly of Stakeholders' Panel on Agent Detection Assays (SPADA) Test Panels. September 13. Final Report. LA-CP 11-01070. Department of Homeland Security (DHS) BioWatch Project. Los Alamos, NM: Los Alamos National Laboratory.

LANL. 2011b. Evaluation of the Critical Reagents Program PCR-Based Assay for Agent 7. November 9. Final Report. Department of Homeland Security (DHS) BioWatch Project. Los Alamos, NM: Los Alamos National Laboratory.

LANL. 2012a. Evaluation of the Critical Reagents Program PCR-Based Assay for Agent 8. November 28. Final Report. Department of Homeland Security (DHS) BioWatch Project. Los Alamos, NM: Los Alamos National Laboratory.

LANL. 2012b. Evaluation of the Critical Reagents Program PCR-Based Assay for Agent 9. November 28. Final Report. Department of Homeland Security (DHS) BioWatch Project. Los Alamos, NM: Los Alamos National Laboratory.

Merlin, T.L. 2014. BioWatch PCR standards—a CDC LRN perspective. Presentation to the Committee on PCR Standards for the BioWatch Program, March 13, Washington, DC.

Merrill, L., J. Dunbar, J. Richardson, and C.R. Kuske. 2006. Composition of *Bacillus* species in aerosols from 11 U.S. cities. *Journal of Forensic Sciences* 51(3):559-565.

Morelli, G., Y. Song, C.J. Mazzoni, M. Eppinger, P. Roumagnac, D.M. Wagner, M. Feldkamp, B. Kusecek, A.J. Vogler, Y. Li, Y. Cui, N.R. Thomson, T. Jombart, R. Leblois, P. Lichtner, L. Rahalison, J.M. Petersen, F. Balloux, P. Keim, T. Wirth, J. Ravel, R. Yang, E. Carniel, and M. Achtman. 2010. *Yersinia pestis* genome sequencing identifies patterns of global phylogenetic diversity. *Nature Genetics* 42(12):1140-1143.

Morse, S.A. 2014. Perspectives on the standards development process. Presentation to the Committee on PCR Standards for the BioWatch Program, June 12, Washington, DC.

Murphy et al. 2012. Strategies for detection of *Plasmodium* species gametocytes. *American Journal of Tropical Medicine and Hygiene* 86(3):383-394.

Naraghi-Arani, P. 2014. Presentation to the Committee on PCR Standards for the BioWatch Program, June 12, Washington, DC.

NRC (National Research Council). 2010. *Sequence-Based Classification of Select Agents: A Brighter Line.* Washington, DC: The National Academies Press.

NRC. 2011. *Review of the Scientific Approaches Used During the FBI's Investigation of the 2001 Anthrax Letters*. Washington, DC: The National Academies Press.

NSC (National Security Council). 2009. National Strategy for Countering Biological Threats. Washington, DC: Executive Office of the President. Available: http://www.whitehouse.gov/sites/default/files/National_Strategy_for_Countering_BioThreats.pdf.

NSTC (National Science and Technology Council). 2011. A National Strategy for CBRNE Standards. Washington, DC: Executive Office of the President Available. http://www.whitehouse.gov/sites/default/files/microsites/ostp/chns_cbrne_standards_final_24_aug_11.pdf.

NSTC. 2013. National Biosurveillance Science and Technology Roadmap. Washington, DC: Office of Science and Technology Policy, Executive Office of the President.

O'Neill, W. 2014. USPS BDS briefing. Presentation to the Committee on PCR Standards for the BioWatch Program, March 13, Washington, DC.

OIE (World Organisation for Animal Health). 2013. Chapter 1.1.5. Principles and methods of validation of diagnostic assays for infectious diseases. In *Manual of Diagnostic Tests and Vaccines for Terrestrial Animals 2014*. Available: http://www.oie.int/international-standard-setting/terrestrial-manual/access-online/ [accessed November 18, 2014].

OMB (Office of Management and Budget). 1998. Federal Participation in the Development and Use of Voluntary Consensus Standards and in Conformity Assessment Activities. OMB Circular A-119.

Omberg K. 2014. BioWatch gen-3 assay evaluation. Presentation to the Committee on PCR Standards for the BioWatch Program, October 1, Washington, DC.

Opel, K.L., D. Chung, and B.R. McCord. 2010. A study of PCR inhibition mechanisms using real time PCR. *Journal of Forensic Science* 55(1):25-33.

Peccia, J., and M. Hernandez. 2006. Incorporating polymerase chain reaction-based identification, population characterization, and quantification of microorganisms into aerosol science: A review. *Atmospheric Environment* 40:3941-3961.

Pillai, S. 2014. PSAA vs FSAPE vs PHAA nucleic acid based detection. Presentation to the Committee on PCR Standards for the BioWatch Program, April 23, Washington, DC.

Quick, J., A.R. Quinlan, and N.J. Loman. 2014. A reference bacterial genome dataset generated on the MinION™ portable single-molecule nanopore sequencer. *Gigascience* 3:22.

Rasko, D.A., P.L. Worsham, T.G. Abshire, S.T. Stanley, J.D. Bannan, M.R. Wilson, R.J. Langham, R.S. Decker, L. Jiang, T.D. Read, A.M. Phillippy, S.L. Salzberg, M. Pop, M.N. Van Ert, L.J. Kenefic, P.S. Keim, C.M. Fraser-Liggett, and J. Ravel. 2011. *Bacillus anthracis* comparative genome analysis in support of the Amerithrax investigation. *Proceedings of the National Academy of Sciences of the United States of America* 108(12):5027-5032.

Schrader, C., A. Schielke, L. Ellerbroek, and R. Johne. 2012. PCR inhibitors—occurrence, properties and removal. *Journal of Applied Microbiology* 113:1014-1026.

Smidansky, E. D., Arnold, J. J., and C. E. Cameron. 2008. Nucleic acid polymerase fidelity and viral population fitness. pp 135-160. In *Origin and Evolution of Viruses* (Domingo, E., Parrish, C. R., and Holland, J. J. eds.), Academic Press, London.

Tettelin, H., V. Masignani, M.J. Cieslewicz, C. Donati, D. Medini, N.L. Ward, S.V. Angiuoli, J. Crabtree, A.L. Jones, A.. S. Durkin, R.T. DeBoy, T.M. Davidsen, M. Mora, M. Scarselli, I. Margarit y Ros, J.D. Peterson, C.R. Hauser, J.P. Sundaram, W.C. Nelson, R. Madupu, L.M. Brinkac, R.J. Dodson, M.J. Rosovitz, S.A. Sullivan, S.C. Daugherty, D.H. Haft, J..Selengut, M.L. Gwinn, L..Zhou, N. Zafar, H. Khouri, D. Radune, G. Dimitrov, K. Watkins, K.J..B. O'Connor, S. Smith, T.R. Utterback, O. White, C.E. Rubens, G. Grandi, L.C. Madoff, D.L. Kasper, J.L. Telford, M.R. Wessels, R. Rappuoli, and C.M. Fraser. 2005. Genome analysis of multiple pathogenic isolates of *Streptococcus agalactiae*: Implications for the microbial "pan-genome." *Proceedings of the National Academy of Sciences of the United States of America* 102(39):13950-13955.

U.S. House of Representatives. 2012a. BioWatch Present and Future: Meeting Mission Needs for Effective Biosurveillance? Serial No. 112-117. Joint Hearing Before the Subcommittee on Emergency Preparedness, Response, and Communications and the Subcommittee on Cybersecurity, Infrastructure Protection, and Security Technologies of the Committee on Homeland Security, House of Representatives, 112th Congress, 2nd Session, September 13. Available: https://www.hsdl.org/?view&did=743066 [accessed October 29, 2014].

U.S. House of Representatives. 2012b. Letter from the Honorable Red Uptan, Chair, Committee on Energy and Commerce, and the Honorable Cliff Stearns, Chair, Subcommittee on Oversight and Investigations, to Dr. Thomas Frieden, Director, Centers for Disease Control and Prevention, November 13.

U.S. House of Representatives. 2012c. Letter from the Honorable Red Uptan, Chair, Committee on Energy and Commerce, and the Honorable Cliff Stearns, Chair, Subcommittee on Oversight and Investigations, to the Honorable Janet Napolitano, Secretary, Department of Homeland Security, November 13.

Walter, M.V. 2014a. BioWatch program overview: Briefing on PCR standards. Presentation to the Committee on PCR Standards for the BioWatch Program, March 13, Washington, DC.

Walter, M.V. 2014b. PCR standards for the BioWatch program. Presentation to the Committee on PCR Standards for the BioWatch Program, June 13, Washington, DC.

White House. 2004. Homeland Security Presidential Directive 10. Biodefense for the 21st Century. Washington, DC: Executive Office of the President.

White House. 2007. Homeland Security Presidential Directive 21. Public Health and Medical Preparedness. Washington, DC: Executive Office of the President.

White House. 2012. National Strategy for Biosurveillance. July 31. Washington, DC: Executive Office of the President.

Wilson, I.G. 1997. Inhibition and facilitation of nucleic acid amplification. *Applied and Environmental Microbiology* 63(10):3741-3751.

Wilson, M.R., S.N. Naccache, E. Samayoa, M. Biagtan, H. Bashir, G. Yu, S.W. Salamat, S. Somasekar, S. Federman, S. Miller, R. Sokolic, E. Garabedian, F. Candotti, R.H. Buckley, K.D. Reed, T.L Meyer, C.M. Seroogy, R. Galloway, S.L. Henderson, J.E. Gern, J.L DeRisi, and C.Y. Chiu. 2014. Actionable diagnosis of neuroleptospirosis by next-generation sequencing. *New England Journal of Medicine* 370(25):2408-2417.

Appendix A

BioWatch Program and Operational Details

Information contained in Appendix A has been determined by the U.S. Department of Homeland Security to be described in Exemption 7 of the Freedom of Information Act, 5 USC § 552 (b)(7)(F). Requests for this document should be directed to the Board on Life Sciences or Office of Program Security of The National Academy of Sciences.

Appendix B
Study Activities

The committee held three information-gathering meetings in Washington, DC, and heard from federal, state, and local officials, medical and laboratory professionals, and academic and private-sector researchers. The committee also heard from presenters in data-gathering sessions that were not open to the public under Subsection 15(b)(3) of the Federal Advisory Committee Act, 5 U.S.C. App. The National Academy of Sciences determined that to open these sessions to the public would have disclosed information protected under the Freedom of Information Act, 5 U.S.C. § 552(b).

The first meeting, held March 13-14, 2014, in Washington, DC, included speakers from the Department of Homeland Security—the sponsor of the study—as well as experts on PCR standards development and evaluation. The meeting objectives were to hear from the study sponsor regarding the statement of task given to the committee, receive an overview of the BioWatch program and its current processes, obtain information about the Stakeholder Panel on Agent Detection Assays (SPADA) PCR standards development process, learn from the Centers for Disease Control and Prevention about their involvement in the BioWatch program and in relevant standards development and evaluation activities, obtain background information on the PCR-based detection system used by the U.S. Postal Service, and hear from public health laboratories on the use of BioWatch assays and from public health leaders in emergency preparedness and response to better understand their issues and needs.

During the second meeting, held June 12-13, 2014, in Washington, DC, the committee learned about the history of federal biosurveillance standards efforts and about relevant activities within the Department of Defense. The committee heard from additional experts involved in developing and validating PCR assays and in the design and development of PCR performance standards. The discussion considered performance characteristics and validation strategies, approaches in the design of PCR performance standards, issues that may affect assay performance and validation, and conceptual and practical tradeoffs between different types of approaches.

At the third meeting, held September 3-4, 2014, in Washington, DC, the committee spoke with additional public health laboratory professionals and public health decision makers within BioWatch jurisdictions on their views and needs with regard to the PCR detection assays and on issues encountered in understanding and interpreting assay results.

The committee also gathered input via several questions online and expresses its appreciation to those who shared their views.

INVITED SPEAKERS

The following individuals were invited speakers at meetings and data-gathering sessions of the committee:

Linda Beck, Ph.D.
Department of Homeland Security

Andrew Cannons, Ph.D.
Florida Department of Public Health

Patrick Cusick, R.S., M.S.P.H.
Cleveland Department of Public Health

Matthew Davenport, Ph.D.
Johns Hopkins University Applied Physics Laboratory

Jeffrey S. Duchin, M.D.
Public Health–Seattle and King County, Washington

Peter Emanuel, Ph.D.
U.S. Army Edgewood Chemical Biological Center

Stella Fogleman, R.N., M.S.N./M.P.H., C.N.S.
Los Angeles County Department of Public Health

Alexander Garza, M.D., M.P.H.
Saint Louis University College for Public Health and Social Justice

Bruce Goodwin
Critical Reagents Program, JPEO-CBD

J. Peyton Hobson, Ph.D.
Food and Drug Administration

Scott Hughes, Ph.D.
New York City Public Health Laboratory

Anne Hultgren, Ph.D.
Department of Homeland Security

Molly Isbell, Ph.D.
Signature Science, LLC

Paul Keim, Ph.D.
Northern Arizona University

Moon Kim, M.D., M.P.H., F.A.A.P.
Los Angeles County Department of Public Health

Luther Lindler, Ph.D.
Armed Forces Health Surveillance Center

Beth Maldin Morgenthau, M.P.H.
New York City Department of Health and Mental Hygiene

Laurene Mascola, M.D., M.P.H.
Los Angeles County Department of Public Health

Suzet M. McKinney, Dr.P.H., M.P.H.
Chicago Department of Public Health

Toby Merlin, M.D.
Centers for Disease Control and Prevention

Jane B. Morrow, Ph.D.
National Institute of Standards and Technology

Stephen A. Morse, Ph.D.
Centers for Disease Control and Prevention

Pejman Naraghi-Arani, Ph.D.
Lawrence Livermore National Laboratory

David Norwood, Ph.D.
U.S. Army Medical Research Institute of Infectious Diseases

Kristin Omberg, Ph.D.
Los Alamos National Laboratory

William O'Neill
U.S. Postal Service

Sally Phillips, R.N., Ph.D.
Department of Homeland Security

Segaran Pillai, Ph.D.
Department of Homeland Security

Benjamin Schwartz, M.D.
Los Angeles County Department of Public Health

Sanjiv Shah, Ph.D.
National Homeland Security Research Center

Colin Stimmler, M.A.
New York City Department of Health and Mental Hygiene

Maureen Sullivan, M.P.H.
Minnesota Department of Health

Robert K. Tran, Ph.D.
Los Angeles County Public Health Laboratory

Michael Walter, Ph.D.
Department of Homeland Security

Appendix C Glossary

Agent: See Biological agent.

Air sampling: Collecting quantities and types of atmospheric contaminants from a representative sample of air.

Amplicon (PCR): A piece of nucleic acid representing a specific product of the amplification reaction.

Assay: A quantitative or qualitative evaluation of the presence or amount of a given target in a particular sample.

Biological agent: A microorganism (or a toxin derived from it) that causes disease in humans, plants, or animals and is used in bioterrorism or biological warfare.

Biosurveillance: The process of active data-gathering with appropriate analysis and interpretation of biosphere data that might relate to disease activity and threats to human or animal health—whether infectious, toxic, metabolic, or otherwise, and regardless of intentional or natural origin—in order to achieve early warning of health threats, early detection of health events, and overall situational awareness of disease activity.

BioWatch Actionable Result: A determination that occurs when analysis of a filter from a BioWatch sampler indicates the confirmed presence of a target organism's nucleic acid signature.

BioWatch jurisdiction: A major metropolitan area—which may include one or more city, county, state, or regional decision-making bodies—where BioWatch air samplers are operational.

BioWatch program: An activity funded by the Department of Homeland Security that uses sets of air samplers in more than 30 jurisdictions to collect airborne particles onto filters that are subsequently transported to laboratories for analysis for the presence of genetic material from certain biological agents.

BioWatch system: The collection of operational components (which are themselves systems) that produce information from air sampling and feed it into a public health decision-making process to determine the appropriate response to a BioWatch Actionable Result (BAR).

Detection: The determination or recognition of the presence of an object or state of interest.

Detector: A data collection and processing technology that both collects and evaluates data.

Environmental background panel: In the context of a biological agent detection assay, a set of organisms that may be present in the environment, against which the detection assay is tested to determine whether the assay potentially cross-reacts with nucleic acid from these organisms.

Environmental sampling: In the context of the BioWatch system, physical sampling of the environment around a collector associated with a positive PCR assay result to provide decision makers with additional situational assessment and to inform appropriate response actions.

Environmental surveillance: Monitoring of the environment to evaluate potential exposure to harmful agents and damage to living organisms.

Exclusivity strain panel: In the context of a biological agent detection assay, a set of organisms such as closely related bacterial or viral strains and species, against which the detection assay is tested to determine whether the assay potentially cross-reacts with nucleic acid from these organisms.

False negative: A negative result for a given target when the target is present.

False positive: A positive result for a given target when the target is not present.

Inclusivity strain panel: In the context of a biological agent detection assay, the set of organisms, such as bacterial or viral strains and species, against which the detection assay is tested to determine whether the assay is successfully able to detect nucleic acid from these organisms.

Laboratory Response Network: A national network of local, state, and federal public health, health care, food, agriculture, veterinary, and environmental testing laboratories that provide the laboratory infrastructure and capacity to respond to biological and chemical terrorism and other public health emergencies.

Limit of detection: The minimum concentration of a substance that an assay can detect above background with a certain degree of statistical confidence probability (frequently with 95 percent confidence).

Monitoring: Periodic or continuous surveillance or testing to determine the presence or level of a substance of interest in various media or in humans, plants, and animals.

Multiplex (PCR assay): A type of PCR assay designed to simultaneously amplify and detect multiple target regions of nucleic acid in a single sample. It is distinguished from procedures that perform one molecular assay at a time (see Singleplex assay).

Nucleic acid signature: Nucleotide sequence that is considered distinctive for an organism, or a subset of organisms, and can be used in testing for the presence of nucleic acid from this organism.

Polymerase chain reaction (PCR): A technique in molecular biology that is used to reproduce enzymatically (amplify) selected sections of DNA. It permits the analysis of any short sequence of DNA (or RNA) without the need for cloning.

Plasmid: A small piece of DNA found within a bacterial cell that is separate from the chromosomal DNA of the organism.

Primer (PCR): A short oligonucleotide that is complementary to part of the target nucleic acid sequence that defines the region that will be selectively amplified by the PCR. During PCR, the primer hybridizes to its complementary sequence and polymerase extends the primer in the amplified region. A pair of primers is used, flanking the target amplification region.

Probe: A short oligonucleotide that is complementary to part of the target region; the probe is labeled in a manner that enables it to detect amplification (e.g., by exploiting fluorescence quenching).

Quality assurance: A comprehensive system and infrastructure that ensures that a program or laboratory meets standards of quality.

Quality control: A process that focuses on the performance of specific testing activities.

Screening: In the context of the BioWatch system, analysis of air sampler filters for a preliminary genetic signature of a particular pathogen.

Sensitivity: The probability that a system will correctly indicate the presence of a particular substance when the substance is present above a certain concentration.

Signature: See Nucleic acid signature.

Singleplex (PCR assay): An assay that is designed to amplify and detect a single target region of nucleic acid per sample. Used in comparison to multiplex PCR assay.

Specificity: The ability to correctly identify the absence of a target substance when it is not present.

Threshold cycle (C_t): In the context of a real-time PCR assay, the cycle number at which the amplification curve crosses a specified threshold level, indicating that amplification of the target region is above a background level. The greater the number of cycles required to cross this threshold, the lower the starting concentration of target nucleic acid in the sample.

Validation: The process of ensuring and documenting that a thing, such as a test, device, or process, fulfills the purpose for which it is intended.

Verification: A process that affirms that a given laboratory can obtain the anticipated results and assay performance from a validated assay.

Appendix D
Biographies of Committee Members

Georges C. Benjamin, M.D., became executive director of the American Public Health Association, the nation's oldest and largest organization of public health professionals, in 2002. Prior to that he was secretary of the Maryland Department of Health and Mental Hygiene, where he played a key role in developing Maryland's bioterrorism plan, following 4 years as the department's deputy secretary for public health services. Dr. Benjamin started his medical career in 1981 in Tacoma, Washington, where he managed a 72,000-patient-visit ambulatory care service as chief of the Acute Illness Clinic at the Madigan Army Medical Center. A few years later, he served as chief of emergency medicine at the Walter Reed Army Medical Center. After leaving the Army, he chaired the Department of Community Health and Ambulatory Care at the District of Columbia General Hospital. He was promoted to acting commissioner for Public Health for the District of Columbia and later directed one of the busiest ambulance services in the nation as interim director of the Emergency Ambulance Bureau of the District of Columbia Fire Department. Dr. Benjamin has served as a member of the Institute of Medicine's (IOM's) Board on Population Health and Public Health Practice and on several other IOM and IOM/National Research Council (NRC) committees in areas such as training physicians for public health careers, measures to enhance the effectiveness of the Centers for Disease Control and Prevention quarantine station expansion plan for U.S. ports of entry, evaluation of the metropolitan medical response systems program, and research and development needs for improved civilian medical response to chemical or biological terrorism incidents. He is currently a member of the NRC's Report Review Committee. Dr. Benjamin is a graduate of the Illinois Institute of Technology and the University of Illinois College of Medicine. He is board certified in internal medicine and is a fellow of the American College of Physicians and a fellow emeritus of the American College of Emergency Physicians.

Kenneth I. Berns, M.D., Ph.D., is distinguished professor emeritus in the Department of Molecular Genetics and Microbiology at the University of Florida. He also serves as a member of the National Science Advisory Board for Biosecurity, a federal advisory committee that addresses issues related to biosecurity. He has served as a member of the Composite Committee of the United States Medical Licensing Examination, chairman of the Association of American Medical Colleges, president of the Association of Medical School Microbiology and Immunology Chairs, president of the American Society for Virology, president of the American Society for Microbiology and vice president of the International Union of Microbiological Societies. Dr. Berns' research examines the molecular basis of replication of the human parvovirus and the adeno-associated virus and its ability to establish latent infections and be reactivated. His work has helped provide the basis for use of this virus as a vector for gene therapy. Dr. Berns received his M.D. and Ph.D. in biochemistry from the Johns Hopkins University. He is a member of the National Academy of Sciences and the Institute of Medicine.

Bruce Budowle, Ph.D., is director of the University of North Texas Health Science Center's Institute of Applied Genetics and a professor in the Department of Molecular and Medical Genetics. He received a Ph.D. in genetics in 1979 from Virginia Polytechnic Institute and State University. From 1979 to 1982, Dr. Budowle was a postdoctoral fellow at the University of Alabama at Birmingham. Working under a National Cancer Institute fellowship, he carried out research predominately on genetic risk factors for such diseases as insulin-dependent diabetes mellitus, melanoma, and acute lymphocytic leukemia. In 1983, Dr. Budowle joined the research unit at the Federal Bureau of Investigation (FBI) to carry out research, development, and validation of methods for forensic biological analyses. Some of Dr. Budowle's efforts over the last decade are in counterterrorism, primarily in efforts involving microbial forensics and bioterrorism. Dr. Budowle is heavily involved in the forensic applications of bioterrorism and has been involved in developing the field known as microbial forensics. In this field, Dr. Budowle has been the chair of the FBI's Scientific Working Group on Microbial Genetics and Forensics, whose mission was to set quality assurance guidelines, develop criteria for biologic and user databases, set criteria for a national repository, and develop forensic genomic applications. He also has served on the Steering Committee for the Colloquium on Microbial Forensics sponsored by American Society of Microbiology and was the organizer of three Microbial Forensics Meetings held at the Banbury Center in the Cold Spring Harbor Laboratory.

Charles Chiu, M.D., Ph.D., is an associate professor in Laboratory Medicine and Medicine, Infectious Diseases at the University of California, San Francisco (UCSF). He is also the director of UCSF-Abbott Viral Diagnostics and Discovery Center at China Basin and associate director of the UCSF Clinical Microbiology Laboratory. Dr. Chiu is an expert in the emerging field of viral metagenomics, and his research is focused on the development of microarray and deep sequencing technologies for viral pathogen discovery and clinical diagnostics. He is also the principal investigator on an R01 grant from the National Institutes of Health on blood bank pathogen screening, California Discovery, UC-MEXUS, and recipient of National Research Fund for Tickborne Diseases grants on the microbial epidemiology of encephalitis, diarrhea, and Lyme disease, a QB3 Rogers Family Foundation Award in translational diagnostics, and a UCSF-Abbott Viral Discovery Award. Dr. Chiu has more than 30 patents and peer-reviewed publications in scientific journals and ongoing collaborations with research groups and public health agencies worldwide, including Abbott Diagnostics, Inc., Global Viral Forecasting, the U.S. Centers for Disease Control and Prevention, the American Red Cross, and the Texas Biomedical Research Institute.

John M. Hardham, Ph.D., is an associate research fellow and technical director for the Emerging Infectious Disease Program at Zoetis, Inc. (formerly Pfizer Animal Health). Dr. Hardham is also a commander in the U.S. Navy Reserve where he serves as the executive officer for the 4th Marine Medical Battalion. After completion of his postdoctoral fellowship at the University of Texas Medical School in Houston, he became a research scientist for Pfizer, Inc. in 1999. Dr. Hardham has served in a variety of roles in the research and development of vaccines and biopharmaceutical products. He accepted a direct commission into the U.S. Naval Reserves as a Lieutenant Junior Grade in 1994. Over the next decade, he served at several naval and fleet hospitals around the country, as environmental health officer and administrative officer, before being mobilized into active duty for the Iraq War in 2003. Dr. Hardham served as the Microbiology Lab director and Preventive Medicine Mobile Medical Augmentation Response Team microbiologist for Navy Environmental Preventative Medicine Unit 6, Pearl Harbor, HI. From 2006 to 2011, Dr. Hardham was assigned to the Office of the Assistant Secretary of Defense for Nuclear and Chemical and Biological Defense Programs, where he served as the deputy medical director and then medical director. In this role, he directed the Department of Defense Medical Countermeasure Development Program and the Advanced Development and Manufacturing Initiative. He served on numerous committees and advisory roles to the Department of Defense and the Department of Health and Human Services and has made briefings to various House and Senate committees, intelligence community programs, executive office committees, and the National Security Council. Dr. Hardham's personal military decorations include the Joint Service Commendation Medal (awarded by Secretary of Defense Robert Gates), the Navy Commendation Award, the Navy Achievement Medal (six awards), the National Defense Service Medal, the Armed Forces Reserve Medal (Bronze M), and the Outstanding Volunteer Service Medal. He received a B.S. in microbiology from The Pennsylvania State University, and M.S. and Ph.D. degrees in microbiology from the University of North Carolina, Chapel Hill.

Grace Kubin, Ph.D., is director of the Laboratory Services Section of the Department of State Health Services in Austin, Texas. In this role, she oversees activities of the Lab Operations and Quality Control Units of the South Texas Health branch, and also serves as the Clinical Laboratory Improvement Act director for both the Austin and South Texas laboratories. She previously served as the laboratory operations unit director (2010-2011) and emergency preparedness branch manager (2007-2010) for the Department of State Health Services. She is a member of the Association of Public Health Laboratories' Public Health Preparedness and Response Committee and has served as a member of the Centers for Disease Control and Prevention/Association of Public Health Laboratories' Laboratory Efficiencies Initiatives Workgroups on Harmonization of Platforms, 2012; Sharing of Test Services Policy Guide, 2012; and Public Health Laboratory Data Management Tool, 2012, as well as Chair of the Laboratory Response Network National Conference Planning Committee, 2012. Dr. Kubin has received honors including the Emerging Leader award from the Association of Public Health Laboratories (2009) and the Charles E. Sweet award for Excellence in Leadership from the Department of State Health Services (2009). She received her M.A. in microbiology and Ph.D. in biological sciences from the University of Texas at Austin.

M. Allen Northrup, Ph.D., is a principal at the Northrup Consulting Group, and is the founder of Microfluidic Systems and a co-founder of Cepheid. He received his Ph.D. in biomedical engineering from the University of California, Davis. He then spent over 8 years as a researcher at the Lawrence Livermore National Laboratory and the University of California at Berkeley, where he was the first to demonstrate the polymerase chain reaction process in a micro-machined silicon chip. He also co-developed microactuators and microactuator materials. Dr. Northrup has 50 issued U.S. and foreign patents, 40 peer-reviewed publications, and is a member of the National Academy of Engineering.

Tom Slezak has been involved with bioinformatics at Lawrence Livermore National Laboratory (LLNL) for 30 years after receiving B.S. and M.S. degrees in computer science from the University of California, Davis. Mr. Slezak is currently the associate program leader for informatics for the Global Security Program efforts at LLNL. He was involved with the Human Genome Program from 1987 to 2000, leading the informatics efforts at LLNL and then the Department of Energy's Joint Genome Institute from 1997 to 2000. In 2000, he began to build a pathogen bioinformatics team at LLNL, pioneering a novel whole-genome analysis approach to DNA signature design. His team developed signature targets for multiple human pathogens that were used at the 2002 Winter Olympic Games under the BASIS program and later adapted for use nationwide in the Department of Homeland Security (DHS) BioWatch program. Under a close collaboration with the Centers for Disease Control and Prevention, the LLNL team has been called on for computational help on smallpox, SARS, monkeypox, avian influenza, and numerous other pathogens. In addition to continuing work on human and agricultural pathogens, Mr. Slezak's team is currently focusing on signatures of mechanisms of virulence, antibiotic resistance, and evidence of genetic engineering. They have been focusing on detecting novel, engineered, and advanced biothreats for several years, leveraging high-risk Intelligence Technology Innovation Center and DHS funding. Mr. Slezak has chaired or served on multiple advisory boards, including the rice genome project, mouse and maize genetics databases, spruce tree genome project (Canada), plant pathogens, and a National Institute of Allergy and Infectious Diseases sequencing center contract renewal.

Peter M. Vallone received his Ph.D. in chemistry from the University of Illinois at Chicago in 1999. In 1999 he was awarded a NRC postdoctoral fellowship that brought him to the Biotechnology Division at the National Institute of Standards and Technology (NIST). After completing his postdoctoral work in 2001, Dr. Vallone became a permanent staff scientist at NIST and he is currently the leader of the Applied Genetics group. During the last 14 years, Dr. Vallone has worked on developing multiplex assays for the detection of genetic variation, has developed methods for the rapid amplification of STR loci, and has been involved in the characterization of nucleic acid–based reference materials. Dr. Vallone has also developed various bioinformatic software tools for the design of nucleic acid-based assays (e.g., AutoDimer). Dr. Vallone has published over 40 peer-reviewed articles in the area of DNA thermodynamics and forensic DNA testing.